Mulberry Sericulture in New Areas
Empowering Tribal Women

About the Author

Dr. Sanjay Kumar Panda did his graduation and post-graduation in Chemistry from the Ravenshaw College, Utkal University, Odisha. He was awarded University gold medals as the best graduate and for topping in post-graduation. He did diploma in forestry from the Indian Forest College, Dehradun.

Dr. Panda joined the Indian Administrative Service in 1980. In course of his working in various capacities in the Governments of Tripura, Odisha and India, Dr. Panda aptly dealt with implementation, supervision as well as policy formulation. His association with promotion of sericulture started while serving as the Director of Industries, Tripura (1988-89). Later, as the Managing Director, Odisha State Tassar and Silk Society and Director of Textiles, Odisha (1989-94), he dealt with implementation of the Bivoltine Sericulture Development Project and the National Sericulture Project in nontraditional area in Odisha. Subsequently, he played a major role in formulation and implementation of the "Tripura Women Sericulture Project"(1994-97). His involvement with promotion of sericulture continued as the Principal Secretary, Planning & Coordination Department, Government of Tripura (2005-09) and later as Chief Secretary (from May 2010 onwards). He has travelled widely within the country and abroad in connection with study on promotion of mulberry sericulture.

Dr. Panda's interest in sericulture relates to its extension in new and nontraditional areas in general and using it as a tool for environmental conservation, poverty alleviation, empowering the people living below poverty line giving attention to the women and the Scheduled Tribe in particular.

Mulberry Sericulture in New Areas

Empowering Tribal Women

S.K. Panda, *IAS*

CONCEPT PUBLISHING COMPANY PVT. LTD.
NEW DELHI-110059

ISBN-13 : 978-93-5125-029-6

First Published 2014

Published and Printed by

Concept Since 1974

Concept Publishing Company Pvt. Ltd.
A/15-16, Commercial Block, Mohan Garden
New Delhi-110059 (India)
T: 25351460, 25351794, *F*: 091-11-25357109
E: publishing@conceptpub.com, W: www.conceptpub.com
Editorial Office: H-13, Bali Nagar, New Delhi-110 015, India.

Cataloging in Publication Data—*Courtesy:* D.K. Agencies (P) Ltd. <docinfo@dkagencies.com>

Panda, Sanjay Kumar, 1955-
Mulberry sericulture in new areas : empowering tribal women / S.K. Panda.
p. cm.
Includes bibliographical references (p.) and index.
ISBN 9789351250296

1. Sericulture—India—Orissa. 2. Sericulture—India—Tripura. 3. Mulberry—Breeding--India—Orissa. 4. Mulberry—Breeding—India—Tripura. 5. Tribes—India—Orissa—Economic conditions. 6. Tribes—India—Tripura—Economic conditions. 7. Women in rural development—India—Orissa. 8. Women in rural development—India—Tripura. I. Title.

DDC 634.380954133 23

सत्यमेव जयते

संतोष कुमार गंगवार
SANTOSH KUMAR GANGWAR

वस्त्र राज्य मंत्री (स्वतन्त्र प्रभार),
संसदीय कार्य, जल संसाधन,
नदी विकास एवं गंगा जीर्णोद्धार राज्य मंत्री
भारत सरकार
नई दिल्ली-110011
Minister of State for Textiles
(Independent Charge),
Minister of State for Parliamentary Affairs, Water Resources,
River Development and Ganga Rejuvenation
Government of India
New Delhi - 110 011

Message

I am extremely happy to go through the manuscript *"Mulberry Sericulture in New Areas—Empowering Tribal Women"* authored by Dr. Sanjay Kumar Panda. The book is based on practical experiences of introduction of mulberry sericulture in new areas outside the traditional mulberry sericulture areas of Karnataka, Andhra Pradesh, Tamil Nadu, Jammu & Kashmir and West Bengal. The author has covered both the technical aspects of mulberry sericulture as well as administrative aspects relating to promotion of sericulture.

The book appears to be a complete and composite guide for all those involved in promotion of sericulture industry in the country in general and in the context of inclusive development for welfare of women and Scheduled Tribes in particular.

I would like to compliment and congratulate Dr. Sanjay Kumar Panda, presently working as Chief Secretary, Government of Tripura, for taking keen interest on the subject and pursuing this consistently over two decades, which is evident from the manuscript. I hope the book will act as a handbook and guide in promotion of sericulture in our country and welfare of the women and tribals in particular.

(Santosh Kumar Gangwar)

केन्द्रीय रेशम बोर्ड
(वस्त्र मंत्रालय भारत सरकार)
केरेबो कांप्लेक्स,
बी.टी.एम. लेआउट, मडिवाला,
बेंगलूर - 560 068

सत्यमेव जयते

CENTRAL SILK BOARD
(MINISTRY OF TEXTILES, GOVT. OF INDIA)
CSB Complex,
B.T.M. Layout, Madiwala,
Bangalore - 560 068.

N.S. BISSE GOWDA, B.Sc., LL.B
Chairperson

Date :06.08.2013

Message

Based on its multiferous advantages and particular relevance to the disadvantaged segments of the society, promotion of sericulture has been taken up by the Ministry of Textiles, Government of India and the Central Silk Board (CSB). Number of plans and programmes are being implemented with a view to improving production and productivity in sericulture as well as the quality of Indian silk for meeting the requirement of silk handloom and power loom industry and reducing dependence on import. Sericulture is a labour intensive activity. With rapid urbanization and industrialization, pressure on land has been increasing and the labour is becoming costlier. As a result, the area under sericulture has been on the decline in the traditional states like Karnataka, Andhra Pradesh and Tamil Nadu. With a view to tackling the situation on the one hand and for extending the benefits of sericulture in to new areas, the CSB has been keenly pursuing extension of mulberry sericulture in to new areas with the help of the concerned state governments. Assistance is being provided to the farmers for raising mulberry, rearing silkworm as well as reeling and twisting of cocoon. Besides, CSB is keenly pursuing research for improving the quality and productivity of Indian sericulture, for making sericulture more profitable with higher return to the rearers. Import of silk yarn from China has also been a matter of concern, which is affecting interest of the domestic rearers. While steps have been taken under the Anti-dumping legislation, effective solution to this problem lies in increasing productivity and quality of Indian silk yarn, which is being attached high priority.

In this context, I am extremely happy to see the manuscript on Mulberry sericulture in new areas — authored by Dr. Sanjay

K. Panda, IAS, presently working as Chief Secretary, Government of Tripura. The author has vast experience of working in sericulture in Tripura and Odisha in various capacities and has traveled widely within the country and abroad. The author has taken keen interest in analyzing various aspects of introduction of mulberry sericulture in new areas, empowering the rearers and providing critical linkages for sustaining sericulture. I am of the firm view that the efforts of the author will be very rewarding for all those dealing with promotion of mulberry sericulture in the country. I would like to compliment Dr. Panda for this painstaking work and hope that the publication will be useful in promotion of sericulture on a sustainable basis in the coming days.

6/8/2013

(N.S. BISSE GOWDA)

RAJ BHAVAN
AGARTALA-799 006
0381-232-4091, 232-5767

DR. D.Y. PATIL
GOVERNOR OF TRIPURA

Foreword

The Constitution of India envisages equality for all. Government has been taking number of steps and implementing plans and programmes for achieving this noble objective. Analysis of development achieved in the past decades indicates that while there has been development, the disparity between regions and persons continues to persist and in certain cases got widened. The twelfth Five Year Plan has, therefore, made "*More inclusive and faster growth*" as the major objective to be achieved. Apart from inclusive growth, the pace of growth is another area of concern. For a country with over a billion population and limited natural resources, it is necessary to evolve a mechanism for ensuring optimum utilization of available resources. It is beyond doubt that the process of economic development has to be made people centric to be done with active involvement of people and people's organisation like three-tier Panchayati Raj Institutions spearheading the process. It is necessary to ensure that various environmental issues are also given due emphasis for making the development process sustainable on a long-term basis. Use of appropriate technology for utilizing available resources optimally and reducing wastage is another aspect that needs specific attention in the development process.

India is a vast country with wide regional variations in people, habit, culture, food habits, occupations, availability of natural resources, etc. Various models of development have been adopted in different parts of the country, some of which have got laurels within the country and abroad. It will be of interest to all those associated with planning and development to analyse the achievements as well as failures of others and draw lessons for replicating the success stories and avoiding failure.

India lives in the villages with large section of people dependent on agriculture and other land based activities. As a result of various steps taken by the Government, the urban areas have been growing fast, however the rural areas continue to lag behind. There are references to the disparities between *"India and Bharat"* and the need for bridging the gap. In this context, agriculture and other land based activities require prime attention. With introduction of new technology, higher investment and technical support, it has become necessary to empower our farmers to pursue various land based activities for achieving higher production and productivity, which will in turn increase their standards of lives. Unlike industrial activities, agricultural operations are constrained by available land and climate. Every region is required to come up with packages of practices suiting the agro-climatic factors.

Out of various land based activities, mulberry sericulture has a distinct place because of multifarious advantages. It is a land based activity which has sizeable employment potential, particularly for the women. It has industrial components as well, relating to reeling, twisting and weaving of silk fabrics. Silk yarn and silk fabrics have a good demand both within and outside the country. Because of these advantages, particularly its strength in transfer of wealth from the consumers, who belong to richer section, to the producers, who belong to poorer section, mulberry sericulture is considered as an effective tool in *establishing socialistic pattern of society* in general and making the process of growth with distributive justice in particular.

In the above background, the book authored by Dr. Sanjay Kumar Panda, covering salient aspects of introduction of mulberry sericulture in new areas, assumes special importance. It has been based on extensive field work done by the author while rendering public service in his capacity as Director of

Textiles in Odisha in the early nineties and later in the northeastern state of Tripura in the early twenty-first century. Strength of the work lies in its application oriented research and specific findings particularly in the context of development of the Scheduled Tribe and women in pursuing mulberry sericulture as a livelihood opportunity. The author has laid specific emphasis on involvement of the primary growers and growers organisation in undertaking developmental work. The criticality of provision of forward and backward linkages has been dealt in the book, which would be of considerable help in introducing the activity in new areas. Covering technical aspects of sericulture, the global scenario and national scenario relating to the silk industry has made the book richer. It is beyond doubt that this would become an ideal reference and companion for the planners, officers and people's representatives in undertaking expansion of mulberry sericulture to new areas. The author, Dr. Panda, deserves sincere commendation for his hard and sincere efforts in bringing about this publication.

Date : 15.03.2013
Place : Agartala.

(D.Y. Patil)

Preface

Rendering public service as a member of the Indian Administrative Service over the last three decades has been a matter of pride and privilege for me. It has provided me a rare opportunity for working closely with the common people and understanding their problems. In course of working in the government, one has come across intentions and interventions for assisting the poor and the underprivileged for achieving the desired objective of "Equality for all". Notwithstanding this intention, eradication of poverty in its totality continues to elude the Nation.

It is beyond doubt that in order to make alleviation of poverty meaningful, the poor man along with his family has to be taken as the focus and all interventions have to be centered on for empowering him in the true sense. In this context, provision of required skill, assets and access to the market assume critical importance. Analysis of reasons behind the success and failure of antipoverty programmes launched from time to time indicates that gaining confidence of the poor men about the objectives, adequacy of the assistance and forward and backward linkages for accessing the market are some of the major constraints. While a number of general remarks and observations have been made from time to time, I feel it would be worthwhile to give specific suggestions for making the antipoverty programmes meaningful. It is in this background that I thought of collating my experience in the field of promotion of mulberry sericulture in the new and non-traditional area.

Mulberry sericulture has several advantages like low capital requirement, good market, being particularly meaningful to women, environmental conservation as well as facilitating transfer of resources from the rich to the poor. It is an ideal antipoverty programme. However, notwithstanding these strengths, mulberry sericulture has remained confined to the five traditional states in India and various attempts for its extension to new areas has met with limited success.

My involvement with mulberry sericulture dates back to 1988, when as the Director of Industries, Tripura, I was entrusted with the responsibility of promoting sericulture along with handloom and handicrafts. Working with the poor men and women with limited assets was a very rewarding experience as it called for innovations and mobilization of the labour without getting constrained by the limited physical infrastructure peculiar to the Northeastern region including Tripura. Subsequently, on an interstate deputation to Odisha, I had the rare privilege of working in this field for a long period of five years (1989-1994) initially as the Managing Director of Orissa State Tassar and Silk Society and later as the Director of Textiles. I was closely associated with implementation of Bivoltine Sericulture Development Project (BSDP), a special project which envisaged rearing of *Bivoltine* (superior Chinese) breed of mulberry silkworm for overcoming the constraints of quality of Indian silk. The project also envisaged socio-economic development of the Lanjia Soura Scheduled Tribe of the Parlakhemundi district of Orissa through mulberry sericulture, which added new dimension to this project. During this period, I had the opportunity of visiting sericulture areas of the traditional states extensively and having close interaction with rearers, reelers, officers, scientists and all other segments associated with industry. I had also the opportunity of studying sericulture in China and Japan, as a member of the Indian delegation under the World Bank sponsored National Sericulture Project.

With a view to pursuing my interest, I had undertaken a socio-economic study of the achievements made under BSDP, for which I was awarded Ph.D degree in economics by Utkal University.

I continued my interest in sericulture after return to Tripura in 1994. Implementation of a women centric special "Tripura Women Sericulture Project" was taken up for promotion of mulberry sericulture in an integrated manner covering mulberry plantation, silkworm rearing, reeling, twisting as well as silk weaving. Six geographical sericulture clusters were developed by organizing the women beneficiaries in to groups and assisting them for pursuing sericulture with rearing of *bivoltine* breed of silkworm. Based on my experience of working in Odisha, steps were taken for development of critical infrastructure viz. *grainage* for production of silkworm seed, facility for training, for collection and marketing of cocoons, *filature* for reeling of cocoon, twisting and weaving of silk fabrics. In brief, sericulture was promoted in an integrated manner from "*Soil to silk*". Subsequently, dovetailing assistance under MGNREGA for raising and maintaining mulberry plantation, preparation of sapling under decentralized people's nursery, construction of rearing houses, etc. proved to be extremely useful in sustaining mulberry sericulture.

Undertaking any new practice particularly with the poor has always been a challenge because of the low income and risk taking capacity. It becomes more difficult in case of a perishable commodity. Introduction of sericulture in a new area becomes particularly difficult as unlike milk and vegetables, the produce cannot be consumed locally. Further, all the three components of sericulture namely mulberry leaves, silkworm and silk cocoon, are perishable. Mulberry leaves are prone to cattle grazing, when it is raised in isolated fields. Mulberry silkworm is prone to diseases particularly in hot and humid climatic condition. Mulberry cocoon is also perishable as the moth emerges within 5 to 6 days of spinning unless it is killed by stifling. Promotion of sericulture in a new area, therefore, requires taking up these activities in a coordinated manner with required technical expertise and close supervision. There were difficulties in converting silk yarn into fabrics due to decline in number of traditional handloom. This was overcome by taking up silk weaving in power-loom and the plain silk fabrics were processed and printed, which fetched a good price.

My journey with mulberry sericulture in new and nontraditional areas over the last two and a half decade has been a process of learning not only about the technical and administrative aspects but more importantly the way the poor men and women think and react to the government interventions. I felt it would be worthwhile to make my experience in the field of promotion of sericulture available to others, who may adopt it in different places with required field specific modifications. This has led to the present publication. I sincerely hope that this work will prove to be useful in promotion of sericulture in new and non-traditional areas.

Dr. Sanjay Kumar Panda
Agartala, Tripura

Acknowledgements

I am deeply indebted to and would like to thank all the concerned officers, staff and primary growers dealing with sericulture in Odisha and Tripura, who have helped me in learning various aspects of sericulture in general and about implementing it at the farmers' fields in particular. I would like to thank Shri T.K. Chakma, Director Handloom, Handicraft and Sericulture, Tripura, Dr. G.B. Singh, Scientist C, Central Silk Board, Shri Pradip Kumar Saha, Sericulture Development Officer, Shri Sarat Kumar Das, Shri Tapan Bhattacharya in particular for the help and support extended in preparing the manuscript. My thanks are also due to Shri Nitin Mittal of Concept Publishing Company Pvt. Ltd. for publishing this work.

I am highly grateful to Dr. D.Y. Patil, His Excellency Former Governor of Tripura for his kind affection, guidance and writing Foreword and to Shri Santosh Kumar Gangwar, Hon'ble Minister of State (Independent Charge), Ministry of Textiles for his Message for this book. I would also thank Shri. N.S. Bisse Gowda, Chairperson, Central Silk Board for his Message.

My sincere thanks are due to Aruna, my wife, for the help and perseverance and Saswati and Aditya, our children, Dr. N.C. Panda, my father for the inspiration and emotional support extended while undertaking this work. On this occasion, I would remember my beloved mother, Shrimati Janaki Devi and pray the Almighty for the peace of the departed soul.

Last, but not the least, I would like to thank the tribal sisters of Tripura, who have taken up sericulture with great zeal and enthusiasm. A recent interaction with Smt. Communistpati Rupini of Bhrigudasbari, Mandai Block, West Tripura on the 27th July 2013 is worth-noting. While recollecting her experience about extension of sericulture in the area, she proudly mentioned that rearing silkworm has become a 'nasha' (intoxication) with the tribal women rearers, who cannot but help rearing silkworm. Her statement reflects the strength and confidence about the success of sericulture in new and non-traditional areas. I feel this is the real strength of India, which each and every person dealing with public service need to appreciate, explore and promote for making poverty a thing of the past.

With great pleasure, I would like to **dedicate this work to Smt. Communistpati Rupini and all other tribal women of Tripura,** who have succeeded in taking up sericulture in a new and non-traditional area.

Contents

List of Figures

Silk Glossary

Antheraea : Genus including famous silkworm species *A.pemyi, A.mylitta, A paphia* and *A.yamamai* producing tasar silk, also includes *A. assama*, yielding muga silk.

Basin waste: This waste consists of cocoons that could not be completely reeled because of too frequent breaks in the thread.

Bave : The thread spun by the silkworm

Boiling off : The process of removing natural gum, also known as degumming

Bourette: Silk noil (in French). It is the coarse yarn spun from carding process and used as raw material. Compared to spun silk it is much thicker and has no luster and is inferior in strength.

Brocade : A woven silk, rich with interwoven design, decorated with gold and silver thread.

Chandrike : A bamboo structure used as mountage for facilitating spinning of cocoon used in India.

Crepe : Means curled or crinkled. Originally it was a silken material; dyed black cloths was heavily sized and treated by a crimping method to give a particular effect and a harsh, granular feel.

Denier : The number of unit weight of 0.05 gm. Per 450 m length of raw silk or continuous filament rayon or synthetic fiber yarn. (Note: Numerically it is equal to the number of g. per 9000 m.)

Double Cocoon : A cocoon jointly built by two silkworms; it serves as a protective shell for both the worms during their pupae stage. Such cocoons are readily distinguished by their size and toughness.

Doupion (Fro or Douppion) : Silk reeled from double cocoons; also known as *Duppione* (in Italian).

Fibroin : The fiber material or raw silk thread that is a protein, not soluble in boiling water.

Filament : An individual strand that is continuous in length. In silk, it may run from 300 to 1400, 1600 and even 1800 yards in length. A fiber or a filament is the smallest unit in any type of fabric. Filaments are finer in diameter than fibers.

Filature : An establishment for the production of raw silk from cocoons.

Floret : The choicest of fibers obtained from spun silk carding.

Floss : Silk from the outer part of the cocoon, low grade and available only in small amounts also known as Blaze (in French). This is not suitable for reeling.

Gum : A protein material that coats the filaments of raw sericin. This gum serves to cement together the two filaments of silk bave and aids in the formation of the cocoon; in silk manufacture, the gum is useful as a natural size in weaving and in the construction of fabrics such as bolting cloth.

Laying : Eggs laid by a moth.

Mulberry : An hardy and perennial tree of genus *Morus*; silkworms (*Bombys mori*) feed on its leaves.

Organdy : A light type of transparent fiber like cotton organdy. It is stiff, has a hard finish, watered or moiré effect and transparency.

Pierced Cocoons : Cocoons from which the moth has emerged after softening the end forcing its way out are not reelable and are used in spun silk.

Raw Silk : Silk as reeled from the cocoon, contains its original gum or sericin.

Rearing House : The house in which silkworms are reared from the hatching of eggs to formation of cocoons.

Reeling : A process of unwinding yarn from cakes or bobbin and rewinding on to a revolving reel.

Renditta: The number of pounds of cocoons required to yield one pound of raw silk.

Satin : Cloth made in satin weave brings one system of threads to the face for the greater part of the time, whereas the other system of yarn will show on the back of the cloth, in an almost solid effect.

Seed Cocoons : Cocoons of pure race and of selected quality (good quality and disease free) chosen for propagation.

Sericulture : The art and science and management of raising silkworms to the production of silk. It includes all activities of silk production.

Silk : A substance that is secreted in the fluid state by arthropods, which on exposure to air hardens in the form of a thread. It is composed mainly of fibroin and partly of sericin and other substances.

Spun Silk : Yarn made from the silk waste and pierced cocoons, by dressing and spinning. The fibers are short, ranging from 2.5 to 20cm. and are spun on the cotton principles.

Throwing : A process of the actual twisting, winding with or without soaking and doubling without drawing of the continuous raw silk fiber to make it into a yarn suitable for knitting, weaving etc.

Unreelable Cocoons : Cocoons which for various reasons being double, dead, diseased, stained and insect-eaten- are unsuitable for reeling.

Waste Silk : The short, unreelable filaments of cocoons including the waste produced during the various processes of silk reeling and throwing.

Wild Silk : Silk furnished by the larvae of wild silkworms; like tasar or tussah silk.

1

Introduction

Making the development process sustainable has been one of the major challenges faced by the mankind in the recent times. Sustainable development[1] refers to a pattern of resource use that aims to meet human needs while preserving the environment so that these needs can be met not only in the present, but also for future generations. It is a pattern which meets the needs of the present without compromising the ability of future generations to meet their own needs. Sustainable development ties together concern for the carrying capacity of natural systems with the social challenges facing humanity. Sustainable development has three parts covering environmental sustainability, economic sustainability and sociopolitical sustainability. The basic concerns, which are interrelated, are two folds *viz*. Environmental relating to checking pollution, carbon emission, global warming and related issues and social relating to reducing the gap between the rich and poor, and eliminating social and economic deprivation. While impact of the environmental issues is felt over a longer term, that of the social is felt on a much shorter term. Improvements in literacy, awareness and spreading of information through Information Technology (IT) based print and electronic media have made the impact of social dimensions felt all over spontaneously in virtually no time.

Environmental sustainability has been under discussion at various national and international forums. The *Rio+20 conferences* held in Rio de Janeiro, Brazil in June 2012 had deliberated on achieving a sustainable future '*the future we want*'. However, the participating developed and the developing countries were at loggerheads over the recommendations. The conference marked a push on major policy shifts to meet the proposed goals with financial assistance from the rich nations. In the context of sustainable development, the conference had recommended steps for eradication of poverty and promoting sustained economic growth, enhancing social inclusion, improving human welfare and creating opportunities for employment and decent work for all, while maintaining the healthy functioning of the Earths ecosystems?

Concept of Sustainable development is being refined by Green development, which seeks to prioritize what its proponents consider to be environmentally sustainable over economic and cultural considerations. Proponents of sustainable development argue that it provides a context in which to improve overall sustainability where cutting edge Green development is unattainable. For example, a cutting edge treatment plant with extremely high maintenance costs may not be sustainable in regions with fewer financial resources. An environmentally ideal plant that is shut down due to bankruptcy is obviously less sustainable than one that is maintainable by the community, even if it is somewhat less effective from an environmental standpoint. Sustainable development is said to set limits on the developing world. The developed countries, which had polluted significantly in the past decades during their own development, are now proposing to set limit for reducing pollution in the developing third world countries, which would impact their economic growth. This is leading to distrust and implementing the recommendations in the true spirit.

On the social deprivation front, agitation and adverse law and order situations have been affecting peace and development in various parts of the world. Reasons for these are often attributed to the unfulfilled aspirations of the youth among other things. One of the recent developments reflecting this sentiment is the "Occupy Wall Street (OWS)"[2] protest movement, which began on September 17, 2011, in Zuccotti Park, Wall Street financial district, New York, USA. It has been widely covered in the global newspaper, Internet or broadcast media. Reducing the social and economic inequality along with greed, corruption and the perceived undue influence of corporations on the government, particularly from the financial services sector were some of the major issues raised by OWS. With a slogan "*We are the 99 per cent*", OWS refers to income inequality and wealth distribution in the U.S. between the one per cent wealthiest and the rest of the 99 per cent population. Significant aspect of this movement relates to the protesters acting on consensus-based decisions made in general assemblies, which emphasized *direct action over petitioning authorities for redress to achieve their goals.*

The strategy for sustainable development adopted so far is required to be revisited keeping in view the above developments. The interventions attempted so far are required to be made more focused for taking care of the aspirations of the youth on the short run without compromising with the long term environmental issues. At the beginning of the new millennium, Millennium Development Goal (MDG) was adopted with a view to achieve certain goals and specific targets by 2015. There is a move for replacing the MDGs with Sustainable Development Goals (SDG) in 2015. This would have norms for green economy, energy efficiency, energy appliances, water efficiency, protecting endangered species such as tigers and whales. The SDGs targets on range of issues including reducing emission for per unit of GDP, preserving bio-diversity, provide quality education, ensure food and clean drinking water to all and gender equality to be met by 2030. Adhering to the MDG, the UN has incorporated some elements of the position taken by India and other developing countries such as that green economy should not result in creation of new green barriers like carbon tax, impose new conditions on aid and finance and increase dependence of the developing countries on rich nations for cleaner technologies.

Indian Scenario

In the Indian context, issues relating to sustainable development are somewhat more complex as it has to deal with multiple constraints typical to a developing country with a high population and relatively low natural resources. As a matter of fact, India has to cater to the needs of about one sixth of the global population with hardly one-sixteenth of the physical resource. While the progress achieved in the last seven decades since independence has been remarkable in several fronts, issues relating to socio-economic deprivation, poverty and unemployment continue to be matter of prime concern. In certain cases, these have been reflected in discontentment among the people in general and the youth in particular. Unfulfilled expectation of the youth is at times leading to frustrations, frustration leading to anger and anger leading to violence. These have in turn been affecting peace and implementation of developmental plans and programmes. This is leading to a vicious circle worsening the conditions of the poor and disadvantaged.

On the environmental front, it appears that while environmental degradation affects all, the poor and the disadvantaged are often the worst affected. As per a study[3] on impact of Climate Change on Women in Agriculture, women farmer and farm labourer bear the brunt of climate change impact. Women at the farm level are doubly disadvantaged. On one hand, their overall workload increases with men migrating out of the village when agriculture production gets affected due to climate change factors. On the other hand, they have poorer access to resources like land, credit and information. In terms of scarcity, it is the women who have less access to food and nutrition due to intra-household inequities. Further during the periods of scarcity, women have to travel longer distances to fetch fuel, water, fodder which exposes them to greater risk. Due to economic as well as social factors, the women are worst hit for example; men migrate out for climatic adaptation whereas women tend to find wage labour locally. This indicates that the environment degradation has strong gender bias against the women.

The challenge posed by developing countries like India is complex and there is no easy answer based on one model which will fit all situations. The solution to the problems shall have to be location-specific and evolved after due consultation with all the stakeholders particularly the people and people's organization. Various alternatives for economic development are required to be examined and opted with location specific requirements for ensuring inclusive growth. Further, the schemes and programmes implemented for employment generation are required to ensure creation of productive assets for checking inflation. In this context, the interest of the disadvantaged segments like scheduled castes, scheduled tribes and minorities are required to be given due attention.

After independence, India adopted the socialist economic model of development with the First Five Year Plans launched in 1950. This model underwent certain basic structural changes based on Liberalization, Privatization and Globalization (LPG) in 1991. The new economic model redefined the role of Government, stressed on liberalization and promoting the private sector (privatization) for facing the challenges emerging as a result of globalization. However, experience of the last two decades under the new economic policy has remained a mixed one. While there are achievements and growths in a number of parameters, it is felt that benefits of growth have not gone equally to all. The distributive aspects of growth and development continue to be a matter of concern. In this context, provision of employment and eradication of poverty have special significance.

In India, agricultural and allied sectors accounted for about 52.1 per cent of the total workforce (in 2009-10). While agriculture has faced stagnation, services have seen a steady growth. Of the total workforce, 7 per cent is in the organized sector, two-thirds of which are in the public sector. Women are mainly employed in agriculture and caring for livestock with only about 20 per cent of the employed women engaged in activities outside agriculture. It has been found that women earn substantially less than men. Unemployment in India is characterized by chronic (disguised) unemployment. The decline in organized employment due to the decreased role of the public sector after liberalization has further underlined the need for focusing on better education and skill formation. Inequalities and inadequacies in the education system have been identified as an obstacle preventing the benefits of increased employment opportunities from reaching all sectors of society.

Eradication of poverty has been another major issue in India. Numbers of plan and programmes have been implemented with the objective of solving the problems of unemployment and poverty by providing financial assistance for setting up businesses, skill honing, setting up public sector enterprises, etc. The strategies adopted in the Food for Work programme and

National Rural Employment Programme is based on using the unemployed to generate productive assets and build rural infrastructure. The major breakthrough in this field has been enactment of the Mahatma Gandhi National Rural Employment Guarantee Act (MGNREGA) in 2005, which guaranteed provision of 100 days of minimum wage employment to every rural household in all the districts of India. Despite its best intentions, MGNREGA is beset with controversy about corruption, deficit financing, poor quality of infrastructure built under this programme, and unintended destructive effect on poverty[4].

Along with these issues relating to environmental sustainability need, Environmental degradation has been accelerating at a fast rate and threatening to reverse the gains on development, particularly the gains made on eradicating poverty. Analysis of the area having higher incidence of poverty indicates that it coincides with areas having high forest coverage. While inadequate physical and social infrastructure is primarily responsible for the backwardness and slow pace of eradication of poverty in the forest areas, there are other factors as well. It is necessary to revisit the strategy for development of forest and forest dwellers for expediting economic development of the backward regions and making the process environmentally sustainable. While undertaking any land based activity, it has to be appreciated that land and water are scarce resources. These are required to be utilized with adoption of appropriate technology in a manner so that the assets are optimally utilized and the environment is protected while giving a decent economic return to the user group. Such approach will make the process sustainable on a long-term basis. In this background, it is necessary to analyze various land based activities for learning from past experiences and designing appropriate strategy for future development.

While tackling these socio-economic problems, the need of the disadvantaged groups like the Scheduled Tribe, Women, etc. are required to be given special attention. Experience of capital intensive ventures with long gestation like hydro-electric power projects taken up in tribal dominated areas indicate that while such projects cause lot of hardship due to shifting and displacement, the real benefit to the displaced tribal families have been minimal. Such poor benefits along with the fear of being swamped by the 'aliens' (non-tribal outsiders) make the tribal skeptic about such ventures. While dealing with welfare of ethnic and tribal people, it is necessary to gain their confidence by reinforcement of indigenous ventures along with capacity building. Prescribing economic activity for the tribal require a genuine helping hand rather than an interfering hand, which would make the economic venture acceptable to the target group and self-sustaining.

In dealing with welfare of the women, access to key social services such as health and education is a critical determinant of the status of women and their ability to participate in the development. The lower attainments of women in key human development indicators are indicative of the sharp disparities in opportunities available to women and men. A worrying factor is the deteriorating child sex ratio. Hence, the key elements for Gender Equity need clubbing the economic development activities with social welfare activities through active participation of the target group. Enterprises of home based workers, self-employed workers and small producers of particular relevance for women are required to be promoted.

Eradication of poverty and unemployment are often attributed to the lack of 'sustainable livelihood'. A livelihood comprises the capabilities, assets (stores, resources, claims and accesses) and activities required for a means of living. A livelihood is sustainable which can cope with and recover from stress and shocks, maintain and enhance its capabilities and assets and provide sustainable livelihood opportunities for the coming generation; and which contributes net benefits to other livelihoods at the local and global levels and in the short and long-term. The rural development has been a major factor of sustainable livelihood by using available resources, manpower and technology. The women and unemployed youths are to be taken as the driving force of any sustainable development in rural India.

Sustainable development has three major aspects. Firstly, in order to make the development sustainable, it is necessary to analyze and plan indicating the role of the major stakeholders on a long-term basis. As natural resources like land, forest, water are limited, it is necessary to use it judiciously. In this context, appropriate technology plays a major role by eliminating wastage and making the use of resources more efficient and viable. With appropriate technology, consumption of natural resources for serving the same purpose gets reduced considerably. For example, use of non-conventional sources of energy like solar energy, wind energy, biogas, etc. reduce the demand for petrol, diesel. The third aspect, which is of paramount importance relates to the consumer and particularly mindset of the consumers. Unless the consumer appreciate the limitation of resources and adverse consequence of consumption led growth, which are not sustainable and are harmful to the humanity in the long run, attempts for generation of more resources and/or adoption of new technology brings only limited relief. For example, simple measures like switching off light, fan and other electrical appliances, when not in use, organizing carpool (sharing of vehicle) for going to the same place, go a long way in meeting one's requirement while optimizing use of scare resources and checking wastage.

As sustainable development depends on a number of interrelated issues, there cannot be any one model for all situations. It is necessary for every region to work out strategy for meeting its needs keeping in view prevailing socio-economic and cultural factors. Sustainable development requires holistic approach, the elected Government is responsible for taking appropriate action for development on a sustainable basis. However, it is necessary for all the other stakeholders to join for ensuring efficient use of scare resources and introduction of appropriate technology.

Based on the experiences of development of last few decades, particularly in the developing countries, it goes without saying that the focus of sustainable development in the coming years shall have to be centered round the people. It is necessary to empower the people by provision of quality health, education and skill development facilities for enabling them to exploit the market. This process by itself would also create new market and expand existing market for goods and services. This would make the development process broad based and socially inclusive.

Relevance of Mulberry Sericulture

The strategy for economic development is required to be drawn up keeping in view the structure of the economy and the state of development. India's large service industry accounts for 57.2 per cent of the country's GDP while the industrial and agricultural sectors contribute 28.6 per cent and 14.6 per cent respectively[5]. Agriculture continue to be the predominant occupation in Rural India, accounting for about 52 per cent of rural employment. The service sector makes up a further 34 per cent, and industrial sector around 14 per cent. While the services sector accounts for bulk of the GDP, its contribution to employment is lower than that of agriculture. Further, it is only the educated and the skilled who can access the services sector, whereas agriculture provides bread and butter to bulk of the poor and unskilled. There lies the importance of the agriculture and allied sector. While provision of quality health and education along with skill holds the key for social inclusion on the long run, in the short-term, strengthening the land based activities appears to be the other alternative. Interests of the small and marginal farmers, the landless labour, most of who belong to socially disadvantaged segments, are required to be given special attention for ensuring balanced development. With a view to increase return to the farmers, it is necessary to adopt new and appropriate technology and evolve suitable cropping pattern depending upon agro climatic factors for maximizing return to the landowner farmers and landless labour. For small landholdings, return from traditional crops like paddy and wheat remains limited. In such cases, crop diversification from paddy to more remunerative crops like vegetables, fruits are required to be pursued with greater thrust.

Mulberry sericulture is a land based activity, which is labour intensive and provides good return to the farmers. It has special significance for women as the silkworm rearing is done indoor, is not physically strenuous and can be taken up with other household chores like taking care of children and the family in cooking cleaning the house, etc. Mulberry sericulture has both agricultural components relating to mulberry plantation as well as industrial components relating to reeling, twisting and weaving. Because of demand of silk yarn and fabrics both within and outside the country, mulberry sericulture has promising market. As a matter of fact, the Indian silk weaving industry is dependent upon import of China silk yarn. Production of quality silk yarn within the country would help in reducing import of silk yarn from outside.

Mulberry sericulture is being practiced as a traditional occupation in certain parts of India for centuries. These include Karnataka and adjoining parts of Andhra Pradesh and Tamil Nadu, Maldah area of West Bengal and the Valley of Kashmir, which are known as the traditional silk growing areas. With a view to exploit potential of mulberry sericulture in generation of employment as well as eradication of poverty, introduction of sericulture have been attempted in several States in the past decades. However, the experiences have remained a mixed one with limited success and more failures without yielding the desired return. As mulberry sericulture is a viable anti-poverty programme, it would be of interest to learn from such experiences while planning for its expansion to new and non-traditional areas.

Notes

1. http://en.wikipedia.org/wiki/Sustainable_development
2. http://en.wikipedia.org/wiki/Occupy_Wall_Street
3. http://www.nrmcindia.co.in/news.php
4. http://en.wikipedia.org/wiki/Economy_of_India#Income_and_consumption
5. http://en.wikipedia.org/wiki/Economy_of_India

2

What is Sericulture?

"Silk" is a fibrous substance produced by some insects, in the form of cocoon or covering, within which the creatures are protected during the period of their metamorphosis from the caterpillar to the moth stage. The webs and nests formed by spiders are also a type of silk. Natural raw silk differs from artificial silk in the sense that later is made by synthetic materials and does not contain nitrogenous matter. The word "silk" was derived from "Si"—the name of the silk worm in Chinese. Silkworm is called "*Soi*" in Korean and "*Seide*" in German.

Sericulture activity has three components. The first component relates to the production of the food plant, of which the silk worm feeds. The second component involves the growing of the silk worm from egg stage to cocoon stage. The third component consists of producing the silk yarn by unwinding the cocoon, popularly known as reeling, followed by further processing like twisting before it is sent for weaving silk fabrics. Each of these three processes is technically intensive in nature. As such, it would be appropriate to briefly discuss the technical aspect of sericulture for having a proper understanding of the issues relating to expansion of sericulture.

Depending on the type of silkworm, four types of silk are produced globally on commercial scale. The mulberry silk moth "*Bombyx mori*" produces mulberry silk, which account for about 98 per cent of the global silk production. Mulberry silkworm rearing originated from China and spread to other countries including India. Other three types of silk are obtained from Tassar silkworm (*Antheraea mylitta*), Muga silkworm (*Antheraea assamensis*) and Eri silkworm (*Philosamia ricini*) respectively.

Latter three types of silk are commonly grouped as "*Non-mulberry silk*". These are also known as "*Vanya silk*", as unlike the mulberry silk worm, which is domesticated, other three silk worms are wild by habits and are reared outdoor. Tassar silkworm grows in wild and feeds on the leaves of the Arjun, Assan and Sal trees. Muga silkworm feeds on the leaves of Som and Soalu trees, which grow in the high rainfall areas of Assam in the northeastern parts of India. Eri silkworm feeds on castor and tapioca leaves. Attempts were made to culture certain varieties of spiders for getting silk. However, the material was found to be unsuitable for weaving. With an output of about 2 per cent of the total world production, non-mulberry silk are of local importance only. The vanya silk are generally coarser and uneven in thickness. Nevertheless, because of their unique nature and peculiarity, non-mulberry silks have their own importance and market.

History of Sericulture

According to historical evidences[1], the silk industry originated from China. Sing Ling-te, the wife of the famous emperor Huang-II, encouraged the cultivation of mulberry tree, rearing of the silk worm and reeling of silk as early as 2640 BC. The empress was personally devoted to promoting the rearing of silkworm. She was also credited with the invention of looms for weaving fabrics. Sericulture drew attention of the royal and noble families right from the beginning. The Chinese guarded the secret of these valuable techniques with utmost vigilance and jealousy, which restricted the spread of the culture to other countries. There are references to imposition of capital punishment on persons, guilty of smuggling silkworm out of China.

The Silkworm rearing technology spread from China to Korea and Japan[2] in the 3rd century A.D. As per *Nihonoi*, the most ancient book of Japanese history, some Koreans were sent from Japan to China to learn the art of weaving and preparing silk goods. They brought four Chinese girls with them, who were instructed to spread the art of plain and figured weaving with silk yarn in Japan. A temple was built in the provinces of Seiidsu in Japan to commemorate these pioneers of silk weaving. Subsequently, considerable efforts were made to promote the industry, which grew in size and played an important role in the

Japanese economy. Historical evidences make it amply clear that mulberry sericulture originated from China and spread to other countries of Asian continent. Silk is believed to be the pioneer of international trade. Being one of the first commodities to be sold across international border, it has lent its name to the ancient trading routes[3] across the Asian and European continents being known as the "**Silk Route**". (Fig. 2.1)

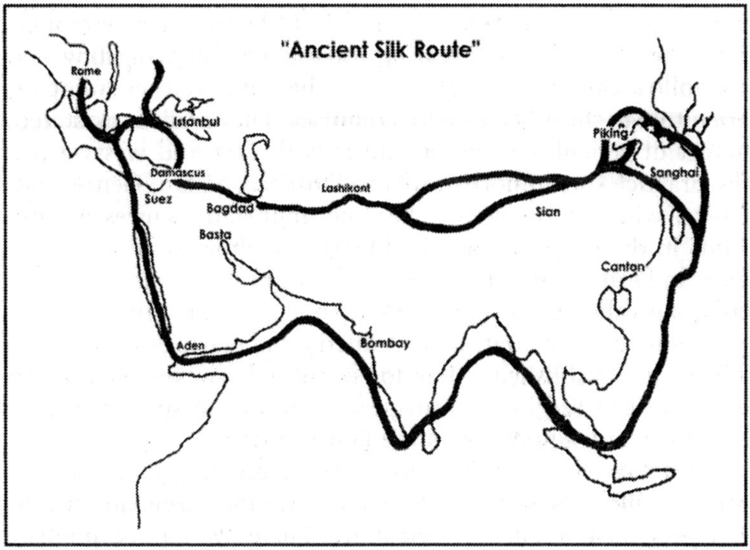

Fig. 2.1: Silk Route

In India, mulberry sericulture was started around 1000 BC. It is believed that eggs of the silkworm and seeds of the mulberry trees were carried to India by a Chinese princess concealed in the lining of her headgear. This is corroborated by early presence of mulberry sericulture activities in the Brahmaputra valley, lying in between the rivers Brahmaputra and the Ganges. From the Ganges valley, mulberry sericulture spread westward to Khotan, Persia and Central Asia. The first authentic reference to silk industry in the western literature occurs in the work of Aristotle "*Hist. Animv*" where he has mentioned of "a great worm—which has horns and so differs from others. Its stages of metamorphosis are egg, caterpillar/larva, pupa and lastly a moth—all these changes taking place within six months. From this animal, women separate and reel off the cocoon and afterwards spin them. It is said that this was first spun in the island of Cos by Ppamphile, daughter of Plato".

Towards the beginning of the Christian era, raw silk became an important and costly items among the products sent from the East to Rome. In an anecdote of Aurelian, silk was said to be worth its weight in gold in the good old days. Importance of silk in trade and commerce grew continuously not withstanding its price and other constraints. Around 550 AD, two Persian monks learnt the art of silkworm rearing and smuggled eggs of silkworm in a hollow cane to Constantinople. This catalyzed spreading of sericulture to the other western countries. The Saracens mastered the art of sericulture and spread it both east and westwards. The practice spread northwards to Florence, Milan, Genoa and Venice, which subsequently became important centres of silk goods in the medieval age. In 1480, Sericulture was started at Tours in France under Louis XI. In 1520, Francis I introduced it in Rhone valley by bringing silkworm eggs from Milan. There was reference to plantation of mulberry and rearing of silkworm in literatures published at London in 1620. However, sericulture could not succeed in the European countries in spite of active assistance and support from the Government.

In 1952, Sericulture was introduced in the American subcontinent in New Spain, Mexico. Later the Virginian settlers were persuaded to take up sericulture. There was a peak of interest in sericulture after the Second World War, when the value of raw silk rose rapidly for making parachutes. There were several speculations on prospects of sericulture industry in America, which did not come true and finally led to uprooting of mulberry. This is reflected in the Compaignie des Indes Occidentalis, "Cultivation of silk occupies a place among the glowing attraction, which allured so many disasters". Even though a good quality of cocoon was produced in America, there were difficulties in reeling. Cocoon was therefore shipped to Japan on experimental basis, which did not succeed. The American silk fabrics manufacturers finally decided to depend on China, the original source of silk to meet their requirement. The Americans assisted the Chinese for developing the sericulture industry by starting three colleges for teaching modern methods of sericulture. The Americans persuaded the Chinese to produce good quality silk so that the American market can be equally shared with the Japanese. However, the Chinese found the American's offer difficult, as it was easier for them to sale the silk prepared by traditional methods to the European buyers, who were not as specific about the standards as the Americans. Subsequently, the American lost interest[4] in this activity following political developments in China. During this period the American had also assisted India in producing silk to meet their domestic demand.

These historical evidences clearly indicate that mulberry sericulture, after originating from China, spread far and wide almost all over the world. Because of its unique texture and lusture, silk was all along considered as a precious commodity and sericulture drew the attention and patronage of the Kings and Monarchs. However, production of silk went through several ups and downs in different parts of the world. In the middle part of the 20th century, Japan was the major producer. Following large-scale industrialization in Japan, production of silk in Japan declined and China became the major producer of silk in the world. From historical facts, it appears that production of raw silk in a profitable manner has been possible in places, where suitable conditions like cheap labour, labour working for long hours at a reasonable wage, high level of efficiency and modern conditions have prevailed.

Technical Aspects of Sericulture

Mulberry silkworm "*Bombyx mori*" belongs to the family *Bombycidae* of the order *Lepidoptera*. The silkworm goes through four stages in its life cycle namely (i) Egg, (ii) larva, (iii) pupae, and (iv) Moth (Fig. 2.2). The moth is white in colour and has a small body measuring about 2 cm in case of the male and a little longer and stouter in case of the female. Its wing is short and weak, the fore pairs are falcate and the hind pairs do not reach the end of the body. The larva is hairless and ash grey to light green colour. The matured larva is slender and 6-7 cm. long. The second thoracic segment is humped and there is a spine like horn at the tail. Under climatic conditions, where the seasons are cold and defined as in Europe, the silkworm produces only one generation in a year. This variety is called "Uni-voltine". In Japan, the reproduction takes place twice a year and the races of silkworm are called "Bivoltine". In tropical climates of India, there are multiple reproductions almost continuously throughout

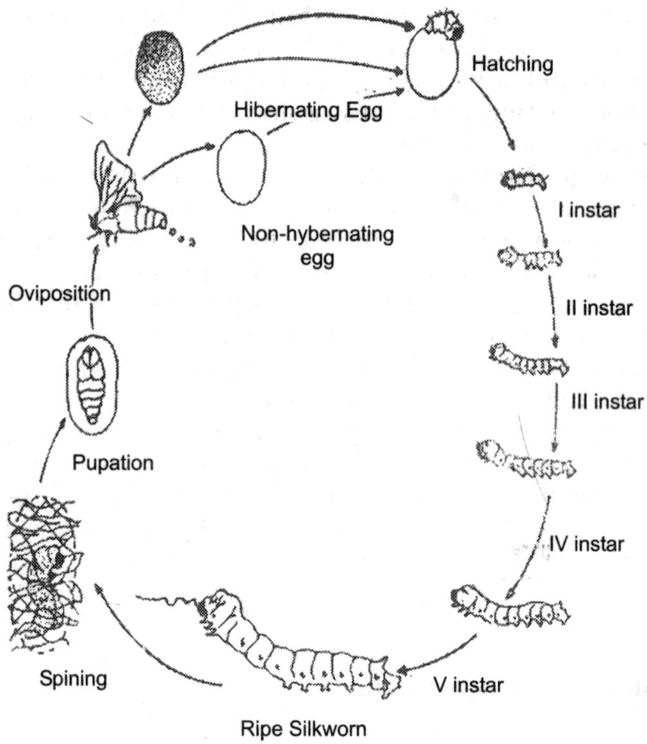

Fig. 2.2: Life Cycle of Silk Worm

the year. These races are called "multivoltine". Quality of the silk usually goes down with increase in voltinism i.e. number of hatching in a year (Fig. 2.3).

Silk yarn is produced by unwinding the filament of the cocoon. The filament is formed out of the fluid excreted from the silk glands of the matured silk worm. The silk glands consist of two long thick walled sacs, running along with the sides of the body of the silkworm. The glands open by a common orifice or seripositor on the under lip of the larva. (Fig. 2.4). As the larva gets matured, these vessels become filled with a clear viscous fluid. This fluid hardens to a solid mass on being exposed to the air. Matured larva spin the cocoon by ejecting the fluid from

Fig. 2.3: Bivoltine and Multivoltine silk cocoons

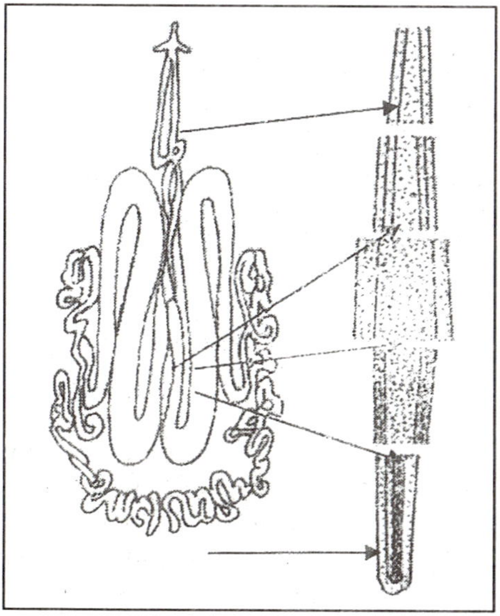

Fig. 2.4: Silk gland of silkworm

both the glands simultaneously, to form a continuous silk thread of about 600 to 1200 meter long moving its head around in a regular order continuously for 2 to 4 days. Simultaneously, a glutinous substance is secreted from two other glands, which hold the thin viscose thread through the spinneret and make the two filaments adhere on coming in contact with the atmosphere. Cocoons vary in size and weight with race of the silk worm. Fully spun cocoon is 2 to 3 cm in length and 1 to 2 cm in diameter. Its colour varies from white to yellow and its shape is oval with or without constriction in the middle. They form hard, firm and compact shells, with some flossy filament on outside. The interior is densely held as a thick mass, which makes unwinding of the thread difficult. The cocoon along with the pupa weighs 1 to 1.8 gm depending upon the breed. After 10 to 12 days of spinning, the moth emerges from the cocoon by softening one end and by pushing out the fibres (Fig. 2.5). The male and female moths mate almost immediately after emergence. After coupling, the female moth lays 300 to 500 eggs at a time, which is commonly known as a "laying". The female moth usually dies after laying eggs, there by completing its life-cycle.

Fig. 2.5: Silk moth emerging from cocoon

Chemical Composition of Silk Yarn

Silk filament consists of two layers of proteins. "*Fibroin*", which in the form of two strands form the inner layer of the silk filament and are surrounded by a kind of gum "*sericin*", as the outer layer. Fibroin ($C_{15}H_{23}N_5O_6$) and sericin ($C_{15}H_{25}N_5O_8$) occur in the ratio of 75:25 respectively. There is a thin layer of albumen and a coat of gelatin outside the sericin. Unlike wool, silk does not contain sulphur. Fibroin is insoluble even in boiling water. The outer layer of sericin is removed by boiling in hot water to make silk soft and loose. Silk is crystalline, homogenous in structure, hygroscopic in nature and light in weight. It is the longest and strongest of all natural fibers. Soft, lustrous and hygienic, Silk has an excellent affinity to dyes. It does not catch fire as quickly as nylon and wool. It has an elastic coefficient of 20 per cent. The breaking strength is about 4 gram per denier. (Denier is the unit referred to the thickness of the filament/yarn. If 9000 m of silk filament/yarn weighs 1 g, then the thickness of the filament/yarn is 1 Denier). Properties of silk as compared to other fibers are given in Table 2.1.

Table 2.1: Properties of Silk Yarn as compared to other Fibres

Fibre	*Specific gravity*	*Hygroscopy*	*Tenacity*	*Elasticity (% gm./dr.)*
Cotton	1.50	8.50	2.5-4.0	7-10
Hemp	1.50	12.00	6.3	1-8
Silk	1.36	11.00	3.0-5.0	16-25
Wool	1.32	16.00	1.2-1.8	30-50
Viscose Rayon	1.55	12.00	1.5-2.4	15-30
Acetate	1.33	6.00	1.5-1.7	23-30
Nylon	1.14	3.40	4.5-5.5	20-25

Production of Silk in Commercial Scale

Mulberry sericulture consists of three components, namely– (i) production of mulberry leaves, known as Moriculture, (ii) rearing of silkworm (indoor under domesticated conditions), known as Silkworm rearing, and (iii) reeling or unwinding of the cocoon to form yarn, which along with other processing is known as post cocoon technology. Silk yarn form the raw materials for weaving of silk fabrics by handloom or power loom.

Moriculture

Mulberry silkworm is *monophagous* in nature and feeds only on the leaves of mulberry "*Morus alba*". Mulberry is a hardy, fast growing deciduous plant. The plant grows well under a wide-ranging agro-climatic condition. In temperate regions, where silk worm is reared only once in a year, mulberry plant is grown in tree form. In case of multiple rearing, mulberry plant is maintained in bush form, which gives more leaves and makes harvesting of leaves easier. Mulberry responds well to agricultural practices and supply of inputs. Leaf yield depends on the quality of soil and efficiency of cultural operations. The bush plantation is pruned periodically at 75 to 80 days intervals. Once planted, mulberry plants yield leaves for many years. The plantations are, however, replaced after 15 years for maintaining vigour and vitality.

Mulberry Plantation

Mulberry plantation is raised from cuttings, saplings, seeds or rootstocks. However, commercial scale plantations are raised from cuttings or saplings. Preparation of a sapling for 5 to 6 months and then using them for plantation yield very good result. Research studies have recommended a number of high yielding mulberry species *Kanva-2 (K-2)*, named after a village near Ramanagaram in Karnataka, is found suitable for irrigated condition. *S-1*, choosen by selection method yields good result under rainfed condition. Certain selected varieties like *S-54, S-34, MR-2, V-1*, etc. reportedly yield higher quantity of leaves. The plantation is usually taken up in June-July with the onset of monsoon. Pencil thick branches, obtained by pruning plants of more than 6 months old, are used for preparation of cuttings for raising plantation. 15-20 cm long and 1 cm thick cuttings, having 3 to 4 buds, are used for raising plantation. Ends of the branches are cut with a sharp instrument in order to avoid any damage to the main stem or peeling off of the bark. Usually 3 cuttings are planted at the corners of an equilateral triangle in each pit. In case of plantation with saplings, one sapling per pit is considered enough.

Before the onset of monsoon, the land selected for mulberry

plantation is ploughed deep and about 20 tonnes of Farm Yard Manure (FYM) is applied per hectare. The manure is mixed well with the soil by ploughing. Thereafter, the land is made into ridge and furrows. Under rainfed condition, pits of 30 cm cube are dug at a spacing of 1 m. by 1 m. Soil, taken out of the pits, is exposed to the sunlight for some time, mixed with FYM, insecticides and then the pit is filled up. In case of irrigated fields, plantation gets established within 5 to 6 months. Under rain fed condition, it takes 9 months to one year for producing enough leaves for a harvest. Plantation raised during June-July yield leaves for taking up a trial rearing during next March. From the next monsoon onwards, the plantation is harvested on regular basis. Plantation raised from saplings gets established much quicker and gives a higher yield (Fig. 2.6).

Fig. 2.6: Mulberry garden

Mulberry leaves are fed to the silkworm either by plucking leaves or by harvesting shoots. The leaf harvest is more labour intensive. Shoot harvest saves labour and time for silk worm rearing. Leaves are usually plucked (from the largest glossy leaf) leaving 3 leaves at the top. The tender leaves are used for rearing young silkworm. The plantation is usually pruned at 20-25 cm. from the ground level with a sharp instrument without damaging the steam or bark. About 30,000 kgs of leaves are harvested in a year from one hectare of mulberry plantation under irrigated conditions. V-1 variety yields about 50-60 MT of leaves per ha. in a year. Under rainfed condition, about 10,000 kg. of leaves are produced per ha. The leaf yield depends on the extent of irrigation, application of manure, chemical fertilizer and other agronomical practices. Organic manure is considered essential for mulberry sericulture. Chemical fertilizers are however applied depending upon the soil nutrient content. Nitrogen, Phosphorous, Potassium are essential for the growth of mulberry. As mulberry leaves are harvested once in every 2-1/2 months, the nutrients get depleted very fast and are required to be replaced. Separate packages of practice have been suggested regarding details of pruning and rearing schedules, for application of organic and chemical fertilizers under rain fed and irrigated conditions.

Diseases of Mulberry

The mulberry plant is susceptible to a number of diseases and pest attacks, of which powdery mildew, leaf spot, leaf rust, root knot diseases and attack by hairy caterpillars are the most common. Powdery mildew usually occurs during the winter and is characterized by the occurrence of white powdery patches on the leaves. The patches gradually turn brown, black and the leaves finally wither off. Spraying fungicides like 0.2 per cent *Carbendazim* can check this disease. Leaf spot occurs during the rainy season, when black spot appear on the leaves. The leaves later turn yellow and become unfit for feeding. Leaf rust disease leads to the appearance of small brownish spots on the lower surface of matured leaves. Spraying of fungicides like 0.2 per cent *Carbendazim* checks these diseases. Root knot usually occurs in mulberry plantation raised in sandy loam soil under irrigated condition. This disease leads to appearance of knot/galls on the roots, which affect yield of leaves. Application of neem oil cake at 400 kg per acre in four equal doses control this disease. Damage by white ants is another typical problem for mulberry plantation, particularly in red laterite soils. This can be controlled by spraying insecticides like gammaxine and the application of neem and karanj oil cakes.

Silk Worm Rearing

The management of the silk worms after hatching from the egg up to the stage of spinning cocoon, form the most important part of the sericulture industry. As a result of many centuries long indoor rearing, the mulberry silkworm, *Bombyx mori,* has become domesticated. The silkworm is reared inside the rearing house, which is maintained exclusively for this purpose. The rearing house is disinfected before and after the rearing for killing harmful pathogens. This helps in the prevention of any disease attack, which affects productivity and revenue from sericulture.

Production of Silkworm Egg

The Silkworm egg is one of the most important inputs of the sericulture industry. It is of critical importance in silk worm rearing. Depending on agro-climatic conditions, pure or hybrid silkworm races are taken up for rearing. Hybrid races offer higher resistance to diseases than the pure races and are preferred for commercial rearing. In India, multivoltine races like Pure Mysore, Nistari were being reared up to the 70's. Subsequently, through initiatives of the Central Silk Board and State Sericulture Departments, rearing of multi-bi hybrid races had been taken up on a large scale. Rearing of pure bivoltine races was stressed for improving the quality of the silk.

Silkworm eggs are prepared in a building having required the facilities, which is known as "*Grainage*". Rearing of parent races [P-1 (first generation), P-2 (second generation) and P-3 (third generation)] which form the mother stock or breeding material, are usually taken up under more hygienic and controlled condition for maintaining the vigour and purity of the races. Seed cocoons are sorted by visual inspection. Deformed or imperfect cocoons, if any, are removed and used for reeling. The sorted seed cocoons are kept in trays for allowing the moth to emerge and mate. After mating, the female moth is separated and kept for laying eggs. Eggs laid by a single moth are called a "*Laying*". After laying of eggs, the female moth is crushed and examined under the microscope for pebrine disease. This is known as moth testing and is of crucial importance in egg preparation. In case any disease is found, the eggs are destroyed by burning. Eggs found to be free from diseases, laid by a single moth is known as one *Disease Free Laying* (DFL). Eggs are disinfected by washing in 2 per cent formalin and supplied to the rearers. In case DFL is not required immediately, it is stored or consigned in the cold storage. Cross breeding of silkworm moths, belonging to two different races, is taken up for having silkworms, which possess characteristics of both the races. Cross breed DFL is usually hardier and better suited for different climatic conditions. It gives higher yield and better quality of silk. In case of preparation of egg of hybrid races, hatching of the two different types of moth are synchronised by keeping seed cocoons in cold storages foı delaying emergence temporarily.

Brushing of Silkworms

Silkworm eggs are prepared either on a piece of paper (egg sheet) or as loose eggs. Under natural conditions, eggs usually hatch at a time, coinciding with budding of mulberry and appearance of fresh green leaves in nature. However, for taking up multiple rearing in a year, the silkworm eggs are processed for expediting hatching and ensuring hatching of all eggs and emergence of all worms of a laying at a time. In the old days, this was being done by keeping the eggs in contact with the human body for warming. Now days, the silkworm egg boxes are kept in the incubator, initially at 18°C and gradually warmed to 25°C for facilitating uniform hatching. Silkworm egg of multivoltine races hatch 9 to 10 days after laying. In case of bivoltine, there is diapause and egg hatch after the winter. Under tropical climate, diapause of egg is shortened by treating the eggs with acid. Acid treatment is done in two ways i.e. either with hot acid or with cold acid. By this technique, it is possible to get hatching of the eggs any time of the year. This is also possible by following the recommended schedule for consignment of layings in the cold storage. Besides rearing under indoor conditions, this is the other major advantage of mulberry sericulture which makes it possible to take up rearing at any time of the year, when adequate quantity of mulberry leaves are available.

After hatching from egg, mulberry silk worm passes through 5 different stages or "*instars*" lasting for about 25 to 30 days (depending on the climate, race and other factors) till maturity i.e. spinning of cocoon stage. During each transition, the worm sheds its scale and passes through a phase called "*Moulting*". During moulting, the worm remains idle and stops feeding for about a day. As the worm grows, its requirement of mulberry leaf increases. As a matter of fact, 92 to 95 per cent of consumption of the leaves occurs during the last two instars. Further as it grows, the worm requires more space in the later instars. Details of the requirement of mulberry leaves and rearing space are in the Table 2.2.

Table 2.2: Requirement of Leaves/Shoots and Space for Rearing 100 DFLs (Shoot Rearing)

Instar	*Leaves/Shoot required (in kg.)*	*Rearing space (Bed Area) (in Sq.Ft)*
First	5	24
Second	20	65
Third	70 Kg Leaves or 140 Kg Shoot	195
Fourth	525 Kg Shoot	380
Fifh	2630 Kg Shoot	700

Silkworm rearing is broadly divided into two parts, (i) rearing of young worm up to 3rd instar which is commonly called as "*chawki worm*" stage, and (ii) rearing of adult worm at 4th and 5th instar stage.

Chawki Rearing

Freshly hatched larvae are usually showered with finely chopped tender mulberry leaves, on which they crawl up and then they are transferred along with the leaves, using a feather, to a rearing tray, inlaid with paraffin paper. This process is called "*brushing*". The tiny ant like worms measure about 3-4 mm each in length and less than 0.5 mg. in weight. The silkworm is a voracious eater and grows very fast. A fully-grown larva weighs about 5g i.e. 10,000 times its weight at the hatching stage, within a period of 25 days, one of the rarest form of growth in living beings.

Rearing of silkworm is usually done in a separate house

known as *rearing house*. In advanced countries like Japan, rearing is done in modern rearing houses, which has facilities for controlling temperature and humidity. However, in developing countries like China and India, rearing is done in pucca, asbestos or thatched roof houses. In some cases, poor rearers, who do not have resources for having a separate house for rearing, take up rearing in part of their living room. In such case, it is not possible to take up disinfection and other recommended practices. Hence, the silkworm becomes more susceptible to diseases as a result of which the quality of cocoon and yield are affected adversely.

Even though *leaf feeding* is more popular, *branch/shoot feeding* is taken up in places where scale of rearing is very high. In the latter case, the whole branches are harvested and put on the rearing shelves in alternate direction. Continuous and close attention is required to be given during rearing for avoiding either wastage of mulberry leaves or malnutrition of the worms. Leaves are harvested either early in the morning or in the evening for feeding the worms. Preserving the harvested leaves in a bamboo basket covered with wet clothes is recommended for keeping the leaves fresh until they are fed to the worms. As the silk worm is sensitive to high humidity, leaf plucking is usually avoided during rainy period. In case of continuous rain, the farmers are advised to wipe out excess water from the leaves with a dry cloth before feeding (Fig. 2.7).

Diseases of Silkworm

Silkworm is susceptible to a number of diseases, which is expected of a creature, propagated for more than 4,000 years under artificial condition. "*Muscardine*", a fatal disease of silkworm, is caused by a fungus, *Beaveria bassiana*. This disease is contagious and infectious. Silkworm, affected by this disease loses appetite, becomes lethargic and ultimately dies. The disease can be checked by regulating humidity and providing proper ventilation. "*Pebrine*" is another infectious and contagious disease, which spreads like wild fire and assumes epidemic proportion. This disease was in fact responsible for wiping out sericulture from France. In case of this disease, skin of the larva gets covered with dark spot, affected eggs either do not hatch or

Fig. 2.7: Silkworm feeding on mulberry leaves

hatch improperly. The worms get weak, stunted and grow unevenly. Cocoons become soft and loose and the moths becomes feeble and inactive. The disease was first discovered by *Pasteur* around 1850 A.D. in France, who suggested that silkworm eggs should be properly tested under microscope for pebrine spores and rearing should be done only after ascertaining that it is free of *pebrine*. This method is being followed extensively to check any outbreak of the disease. As a matter of fact, every single laying supplied to the rearers is tested for pebrine and certified to be *Disease Free Laying* (DFL) before it is supplied to the rearers. This method has been responsible for sustaining sericulture production in the world to a great extent. *Pebrine* disease had been afflicting the Indian silk industry once in every 4 to 5 years. Maintenance of hygienic condition in the grainages, rearing house and regular testing of female moth are essential for checking this disease. "*Grasserie*", another important disease of silkworm, usually occur in the 4th or 5th stage before or after the moulting. Skin of the affected worm gets swollen and thin. This is characterised by decomposition of blood to a milky fluid, which comes out of the skin if ruptured. This disease is viral and assumes dangerous magnitude. Feeding of silkworm with good quality leaves, maintenance of proper ventilation and spacing helps in preventing this disease. "*Flacherie*" is another contagious disease of the silk worm, which occurs due to high temperature and bad ventilation. It can be checked by maintaining proper humidity and temperature in the rearing room.

Spinning of Cocoon

As the silkworm gets matured, it stops feeding and looks for support or anchor for spinning the cocoon. This is called *mounting*. Non-mulberry silkworm usually mounts on the leaf and stems and spin the cocoon. In case of mulberry silkworms, artificial mountage are provided for facilitating the matured worm to spin. The bamboo structure with coiled strips, popularly known as "*Chandrika*" are used because of the low cost and easy availability. Besides mountages made of plastic, and rotary mountage are also used, which cost more but are more durable and easier to maintain. Rearers usually pick up the matured silkworm and place them in the mountage. The worm first forms an anchorage on to the mountage and then moves it head, spinning the cocoon continuously for about 3 to 4 days. The caterpillar gets metamorphosed to a pupa, inside the cocoon. In about 12 days, the pupa gets metamorphosed into moth and emerges out from the cocoon by piercing one end of the cocoon.

After spinning is completed, cocoons are removed from the *mountage*. Good quality seed cocoons are selected for production of eggs. Other cocoons are separated and used for extraction of silk yarn, which is commonly known as "*reeling*". In commercial rearing all the cocoons are used for reeling. The silk worm usually spin cocoon with continuous thread in a manner forming figure "8". As such when the silk moth emerges from the cocoon, the continuous silk thread of the cocoon gets ruptured at many places and the pierced cocoon become unfit for reeling and practically valueless. The pupa is, therefore, killed without damaging the shell of the cocoon to prevent transformation of the pupa to the moth and its emergence from the cocoon. The pupa is usually killed by suffocation without causing any harm to the silk thread. It is popularly known as "*stifling*". Stifling by exposing the cocoons to hot sunlight is commonly followed in rural areas, however, this method affect the quality of silk yarn reeled and is not feasible during rainy season. Stifling by hot sunlight hardens the gum and makes unwinding difficult. It leads to higher wastage and affect quality of yarn. Stifling of cocoon by using electrically operated hot air chambers or dryers gives the good result. However, this is costlier, both on account of the cost of the machine and the cost of power, and is not feasible in interior areas where supply of electricity is erratic. For stifling cocoon, Chinese have developed small constructions, having arrangement for circulating hot air produced by burning mulberry twigs and firewood. These are popularly called" *Ushna koti*", have been found to be cheap and convenient and a case of appropriate technology in rural areas (Fig. 2.8).

Fig. 2.8: Rearer with Freshly Harvested Cocoon

Marketing of Cocoon and Reeling

Last but not the least important part of sericulture industry involves unwinding of the silk cocoon to get silk yarn or thread. This operation is called "*reeling*". Freshly harvested cocoons are either reeled immediately (before emergence of the moth) or are stifled (by killing the pupae inside for avoiding emergence) and stored temporarily before reeling. In case reeling is delayed, stifled cocoons are stored in a godown, having arrangement for maintaining the temperature and the humidity for preventing any decomposition or damage.

Reeling of cocoon to yarn is done by several devices, which are operated either mechanically or by power. The hand operated reeling machines are of two types– (i) the primitive reel called *charkha,* and (ii) the improved type of appliance called *cottage basin.* In some areas of India, charkhas are used with improvements for reeling second grade cocoons and for imparting training to the new reelers. The cottage basins have two separate parts, for cooking and reeling. The multi-end basins are advanced devices, which has arrangements for maintaining evenness of the thread. Fully automatic reeling machines are used for reeling cocoon in Japan, which produces silk yarn of high quality and uniformity (Fig. 2.9).

Fig. 2.9: Silk yarn

Freshly harvested or stifled cocoons are cooked in hot water for softening by removal of sericin before taking up reeling. Depending upon the type of cocoons and reeling, cooking is recommended either with a single pan or three pan. In case of bivoltine races, pressurised cooking is taken up. After taking out bulk of silk from a cocoon, a thin layer of cocoon remains, which is not reelable. This is called "*pelade layer*". Reeling waste is used for making *spun silk.* Silkworm pupa, which is left in side the cocoon, is a rich source of fat and protein. This is used for feeding poultry birds. After the silk yarn is reeled and dried, it is usually wetted with a surface-active compound to check breakage and re-reeled for making it smooth and continuous. Re-reeled raw is laced and then it is mechanically turned into "*skein*", weighing 70 to 80 gram about 30 skeins are bound together in one *book* and are transported for use in weaving. Depending upon the need, the silk yarn is *twisted* in to 2 or 3 plies, which give it additional strength, particularly for use as warp. Silkyarn is woven in handloom and powerloom for making silk fabrics, which is sold after printing and processing (Fig. 2.10).

Fig. 2.10: Silk fabrics for sale

Notes

1. Nanavathy, M.M., *Silk Production and Processing and Marketing*, Willey Eastern Limited, new Delhi, 1990, p. 1.
2. *ibid.*
3. Hyde Nina, The Queen of Textiles, *National Geographic*, January 1984, p. 13.
4. Nanavathy, M.M., *op. cit.*

3

Sericulture: Global Scenario

As per historical evidence, production of silk yarn and weaving of silk fabric originated from ancient China. In due course, it spread to other countries via the '*Silk Route*' and became popular among the rich and the high in the society. Initially it was mainly used as traditional wear for ceremonial occasions. Over years, it came within the reach of the middle class. However, silk accounts for a small percentage, 0.2 per cent of the global textile[1]. Man Made Fibre (MMF) and Cotton account for 65.3 per cent and 32 per cent respectively. Bulk of the mulberry silk (about 90%) and entire non-mulberry silk is produced in the Asian continent.

Sericulture is a land based labour-intensive activity. Its promotion depends on availability of land and labour. Mulberry sericulture is practiced in the developed as well as developing countries. A global survey of the industry reveals that mulberry sericulture is practiced in as many as 29 countries namely China, India, Japan, The Republic of Korea, the erstwhile U.S.S.R., Brazil, Italy, France, Spain, Greece, Yugoslavia, Turkey, Sri Lanka, Hungary, Egypt, Syria, Poland, Bulgaria, Iran, Lebanon, Thailand, Burma, Bangladesh, Afghanistan, Cyprus, The Republic of South Vietnam, Indonesia, Rumania and Cambodia. In the developing countries, sericulture is practiced mainly as a small-scale, village based, labour intensive activity for providing employment to small and marginal farmers and landless population. On the contrary, in the developed countries like Japan, sericulture is practiced on a larger scale as an agro-based industrial venture.

China is the world's biggest producer and main supplier of silk to the world markets. About one million workers are employed in the silk sector in China. India is the world's second largest producer with over 700,000 households deriving their livelihood from this industry.

Because of its multifarious advantages such as high employment potential, rural base, relatively low capital requirement, checking migration to urban area, meeting raw material needs of the silk weaving industry, introduction of sericulture have been attempted in several countries. Even though attempts were made for introduction of sericulture in several countries, it could be sustained only in a few countries. An account of raw silk production in the major silk producing countries is given in the Table 3.1.

Table 3.1: World Raw Silk Production

(*Silk Production in 000 MTs*)

Country	*1990*	*1996*	*2000*	*2005*	*2009*	*2010*	*2011*
China	46.40	68.50	59.800	105.360	104.000	115.000	104.000
India	11.48	14.12	15.850	17.305	19.690	20.400	23.060
Japan	5.72	02.58	0.550	0.150	0.090	0.053	0.044
South Korea	1.20	00.14	0.015	0.150	0.135	0.135	0.135
Brazil	1.69	02.27	1.380	1.280	0.810	0.770	0.558
Others	6.37	08.27	4.250	4.600	2.260	3.692	3.673
Total	**72.86**	**95.88**	**81.845**	**128.845**	**126.985**	**140.050**	**131.470**

Source: ISC web-site

Note: Figures of India is for financial year April to March.

World silk production increased from 72.86 (000MT) to 128 (000MT) in 2005 and to 140 (000MT) in 2010. However, the growth has remained uneven since 2005. China has been leading the world silk production with 79.1 per cent about 4/5th of the total production. India is the second largest producer with 17.5 per cent share of global raw silk production. However, silk productions in most of the countries have remained uneven.

Analysis of production of silk over past decades indicate that China's share in the world market increased steadily from 8.5 per cent in 1938 to over 51 per cent in 1982, to as high as 73 per cent in 2000 and going up to 79.1 per cent during 2011. During the said period, share of India in global silk production increased from 0.7 per cent in 1938 to 9.9 per cent in 1982, 19.4 per cent in 2000 and came down to 17.5 per cent during 2011. During the period, production of silk in Japan fell steeply both in absolute and relative terms. Share of Japan in world silk production fell from 76.3 per cent in 1938 to 25 per cent in 1982 and became marginal, limited to only 0.67 per cent in 2000. This further declined to a negligible 0.03 per cent during 2011.

So far as quality and productivity of silk is concerned, by adoption of advanced scientific technology, Japanese farmers have been able to get about 15,000 kg of mulberry leaves per hector, which led to production of about 712 kg of cocoon and up to 117 kg of raw silk per hector of mulberry plantation. This is much above the productivity level in other countries like India, with production of up to 93 kg raw silk per hector per annum. Economy of the industry measured in terms of cost of production of silk yarn has been guided by five main factors namely efficiency of leaf production, utilization of leaves by the silk worm races, spinning quality of silk, reeling and twisting industry. Out of these the first two activities namely production of mulberry leaf and rearing of silkworm for production of silk cocoon are dependent on manual labour. Therefore, availability and cost of labour play important role in sustenance of sericulture operations.

Keeping in view the demand for raw silk, steps were taken by FAO in the recent decade to introduce Sericulture in Africa and Latin America. While this led to success especially in Brazil and Columbia, it did not make much headway in other countries of the European Union. The European markets continue to depend on imported raw silk from outside Europe for its silk weaving industry.

Analysis of the trend of silk production indicates the following broad features.

- Mulberry silk continues to be the most important silk produced globally.
- China continues to dominate global silk production.
- India has replaced Japan as the second largest silk producer.
- With industrialization, production of mulberry silk has been decreasing.

Status of Global Silk Industry

Even though Silk constitutes only a miniscule percentage (about 0.2 %) of the global textile fibre, it has its own significance in the international trade. Actual trading value of silk and silk products has been quite impressive. As unit price for raw silk is about twenty times that of raw cotton, international silk trade has been a multibillion dollar. The turnover of the China National Silk Import and Export Corporation alone is about 2.5 billion US$ per annum. India, despite being the second largest producer, continues to be one of the largest importers of raw silk.

Italy has been traditionally the largest importer, processor and exporter of silk products in Europe. In 1997, Italy imported about 3200 tonnes of raw silk and over 700 tonnes of silk yarn, primarily from China. Italy also imported about 300 tonnes of silk garments mainly ladies' blouses, of which over 80 per cent came from China. Silk garment imports, however, have drastically gone down over the last five years. Italy is well-known for highly developed skills in processing, finishing, dyeing and printing silk fabrics. During 1996-97, exports of silk scarves rose by about 15 per cent to 586 MTs and exports of silk neckties went up to 1230 MTs.

France has a strong silk processing industry. Lyon is famous for production of high quality silk fabrics for meeting its domestic consumption as well as export. More than 70 per cent of silk fabrics in the French market have been traditionally used for clothing. Use of silk may for interior decoration as curtains, wall covers, bed spreads and upholstery etc. has been increasing. France exports top quality silk fabrics to the US market, with unit prices reaching US$ 30 per m^2.

The US market for silk fabrics including garments, interior decoration fabrics and accessories is one of the worlds largest. As there is hardly any silk processing capacity, the demand is met by import. About US$ 2 billion worth silk goods including 10 per cent for home furnishing were imported during 1997. Unlike European consumers, US consumers do not have a long tradition of using silk. Silk therefore, has never had the same aura as in Europe. The United States has been a pioneer market for imported Chinese knitted silk products. The product range was mainly thermal underwear initially which has widened covering elegant casuals in the form of T-shirts, polo neck sweaters, etc. Keeping in view the preference for easy care, fabrics with easy-care properties are developed to compete with other fibers in the United States market.

In view of the preference for natural fabrics, Germany is one of the largest European markets for textiles and clothing, including silk. The market is quality-conscious and prepared to pay a premium for good quality. A wide range of silk garments, accessories, cushion covers and interior decoration fabrics are imported by Germany. Silk garments are imported mainly from China. Home furnishing materials made in handloom in India and Thailand has been successful in this market.

Japan has been one of the largest silk consumers for meeting its demand for traditional ware mostly for kimonos. It has the highest consumption of silk per capita in the world. It is met entirely on local silk production till mid Sixty's. As domestic silk production of Japan declined sharply after 1970s, it had to depend on imported silk goods, particularly from China. The decline of the Japanese silk processing industry has been affecting sericulture industry of Brazil as it was mainly catering to Japan. Italy is the world's largest importer of waste silk which is spun into yarn for the production of knitwear. China exports the largest quantities of silk. USA imports the largest quantities of silk textile and clothes. Traditionally Asia was producing large quantities of silk and Europe was second.

Silk Export and Import

Operations relating to production of mulberry leaf, rearing of silkworm and production of cocoon are land and labour intensive in nature. On the other hand post cocoon stage operations like reeling, twisting, weaving and processing of silk fabrics are industrial activities, which do not require as much land and labour. As a result, the centers for processing of silk yarn to fabrics have come up at places different from those of the silk cocoon producing areas. Silk yarn is therefore, exported from areas of production to areas of silk weaving. Silk fabrics, which are consumed all over the world, are exported as per the taste and demand. An account of silk yarn imported and exported by different countries is given in the Table 3.2

Table 3.2: Import and Export of Raw Silk by Different Countries

(*In MT*)

Country	YEARS							
	2008		2009		2010		2011	
	Export	Import	Export	Import	Export	Import	Export	Import
Algeria	0	18	0	0	0	0	0	0
Argentina	0	5	0	0	0	0	0	0
Brazil	0	125	728	76	607	0	374	70
Bulgaria	0	0	0	30	0	9	0	32
China	12688	203	8504	173	12900	177	6697	69
Egypt	0	0	0	0	15	98	0	4
France	0	0	7	111	0	108	0	0
India	53	7282	308	6539	154	5159	0	3025
Italy	0	1040	196	501	357	699	362	1860
Japan	0	902	0	725	0	732	0	559
Pakistan	0	0	0	0	0	23	2	0
Philippines	0	20	0	0	0	0	0	0
Romania	0	1	0	0	0	0	0	0
Switzerland	0	35	0	0	0	5	0	0
Syria	0	0	0	0	0	0	0	0
Thailand	0	50	3	30	8	27	0	0
Tunisia	0	68	0	47	0	0	0	0
Turkey	0	100	9	55	19	93	0	0
Uzbekistan	568	0	1710	0	0	0	0	0
Total	**13309**	**9849**	**11465**	**8287**	**14060**	**7130**	**7435**	**5619**

Source: Central Silk board.

From the data, it appears that China enjoys the lion share (about 90%) of the global raw silk export. Raw silk export from all other countries is very meager. Although India is another leading silk producing country after China, its share in export of raw silk is meager (less than 3%). This is because of the fact that bulk of silk produced in India is of multi-voltine nature and is not suitable for use in power looms. Bulk of Indian silk is used in the handlooms for production of sarees the traditional Indian women apparel. India imports sizable quantity of silk yarn for meeting its domestic need for quality silk. Italy and Japan are the other major raw silk importing countries. More than 25 countries have been importing raw silk regularly, which indicates the prevailing demand for quality raw silk in the world. There has been a gap between demand and supply of quality raw silk, which indicates the scope producing quality raw silk in the coming years. Global trade in silk and silk products can be broadly classified under three types of trading countries.[2]

Producers

The first category includes countries, which are major producers and exporters of raw silk and silk waste. China is the largest exporter of raw silk (not thrown) having a share of 92.8 per cent (13,431.1 MT) in total world exports. Romania, Italy and India are the other major exporters with shares of 1.7 per cent (239.5 MT), 1.1 per cent (216.6 MT), and 1.1 per cent (216.6 MT), respectively.

China is the largest exporter of silk waste as well, with a share of 59.7 per cent (2,227.1 MT) in total exports. Germany, United Kingdom, Italy are the other major exporters with shares of 10.8 per cent (403.4 MT), 3.1 per cent (114.1 MT), and 3.0 per cent (112.6 MT), respectively.

Processors

The second category includes countries which are not necessarily producing silk yarn but are major processors but due to strong technological and other expertise.

China is the largest exporter of woven fabric of silk with a share of 23.3 per cent (5,177.9 MT) in total exports.

Hong Kong SAR, Italy and India are the other major exporters of woven fabric of silk having a share of 15.1 per cent (3,348.5 MT), 12.4 per cent (2,749.7 MT) and 10.5 per cent (2,336.2 MT) respectively. Italy has been traditionally the largest importer, processor and exporter of silk products in Europe.

Italy is the largest exporter of Shawls, scarves etc. of silk with a share of 50.2 per cent (2,343.1 MT) in total exports. China, India and France are the other major exporters of shawls, scarves, etc. of silk, having share of 7.4 per cent (343.9 MT), 7.2 per cent (337.7 MT) and 7.1 per cent (329.4 MT) respectively.

Consumers

Consumers who are major importers of silk products and/or are major importers of silk yarn and not major exporters of silk products.

China, Hong Kong SAR is the largest importer of woven fabric of silk or of silk waste with a share of 11.6 per cent (3,874 MT) in total imports. Italy, USA and India are the other major importers of woven fabric silk or of silk waste having a share of 10.3 per cent (3,459.8 MT), 7.9 per cent (2,649.5 MT) and 4.6 per cent (1,555.6 MT) respectively.

The General Agreement of Trade and Tariff (GATT) envisages reducing barriers and facilitating free flow of goods across national boundary. However, the cost of production of silk yarn and fabrics depends on a number of factors internal to each country. There are instances of dumping of silk yarn, which affect the domestic silk industry. India had levied anti-dumping Duty on the low grade silk yarn and fabrics imported from China to support the prices of sericulture commodities in the domestic market. While anti-dumping provisions help in protecting the domestic industry, it is only a short term measure and is required to be accompanied by introduction of technology and measures for improving quality and productivity of silk industry.

Future of Global Silk Industry

Demand for natural fibre like cotton and silk is likely to increase in the coming days due to increase consciousness about health related advantages of natural fibre over manmade fibre. General economic development and increase in individual income has been providing a boost to demand for silk fabrics. The wave of sand-washed silk fabrics seems to have ebbed in Europe and USA and several silk lovers have assigned this development as vulgarization and condemned the flood of cheap silk garments and damaging the queenly status of silk. However, such fears are not likely to lower the prestigious position of silk. Quality of silk will continue to be an important factor in the coming years. On the whole, the future of silk appears to be bright in the twenty-first century.

Production of natural fabrics like cotton and silk are land based activities. Availability of land, which is a basic prerequisite for promotion of sericulture industry, has been shrinking rapidly due to rapid urbanization, industrialization and development of road and rail infrastructure related requirement of land. This would affect growth of the silk industry in the coming decades. The worldwide natural fibre survey[3] had predicted that future global growth impact of natural fibre will be dependent on technical breakthrough, perhaps from genetic engineering which will lead to increase in productivity of high yield per acre of land. Such technological development would reduce cost of production and make the natural fibre competitive as compared with synthetic fibers. Because of climatic factors, countries of the Asian continent will continue to be the main plank of silk production. The leading silk producer, China, has however been losing its coverage under mulberry and cocoon production continuously. In Japan, this has already decreased considerably. In the Indian context, the position has remained more or less static. In this background, new and non-traditional area will provide an opportunity for expansion of sericulture.

Global demand for silk is likely to increase in the coming years on account of the five factors as indicated below.

(i) *New Concept*: Until 1980s silk was worn only for special occasions. But silk is no more a luxury or ceremonial dress. It is now a casual wear, and even sports-wear.

(ii) *New products*: The sand-washed garments which were popular during mid 1990s have lost their share in the market. However, a range of silk products including thermal wear, sportswear, undergarments, shirting and crease-resistant fabrics have come up in the market.

(iii) *New Prices*: Initially Silk was confined to a small affluent minority and was out of the reach of the common man. However, due to new industrial production methods and economy of scale, combined with low wages in many developing countries, prices of silk have come down to within the reach of the middle class during 1990s.

(iv) *New Channels of Distributions*: Silk was initially an exclusive product confined to the high-priced boutiques. It is becoming more visible and touchable on the shelves of the super markets and departmental stores; and

(v) *New Consumers*: Silk was initially confined mostly to women's wear and men could expect at the most a silk tie. A range of product for both men and women have come up covering shirts, jerkins, socks, T-shirts, shorts and few other items.

The wave of sand-washed silk garments seems to have ebbed in Europe and USA and several silk lovers did call this development as 'vulgarization' and condemned the flood of cheap silk garments as damaging the queenly status of silk. Even some purists also condemn the growing concept of silk blends. But such fears are not likely to lower the prestigious position of silk. "It is the quality of the product which will make people continue to buy silk", according to ISA, which feels quality can be guaranteed with adoption of ISO 9000 standards, which are designed to offer the consumer guaranteed performance and quality.

Based on an international treaty, the International Sericulture Commission (ISC) was instituted with its head office in Lyon in France during 1960. The ISC aims at promoting the global sericulture and silk industry. The ISC has been instrumental in the development of silk industry in many countries, including India. At present, there are 13 countries in ISC including Brazil, Egypt, France, Greece, India, Indonesia, Iran, Japan, Syria, Madagascar, Romania, Thailand and Tunisia.

The purity and sanctity of silk is required to be ensured. The silk mark has been introduced as per the recommendation of the International Silk Organization which has examined the strategy for promotion of silk with top fashion designers and large departmental stores marketing the high price silk fabrics with silk mark which guarantees purity of silk. This is also part of the UNCTAD-GATT general agreement on trade and tariff guiding global trade on silk fabric.

Keeping in view the advantages of silk in promoting inclusive development, promotion of silk will continue to be an important factor in the coming days. Many developing countries in Asian and African regions can emulate the development of silk industry in India as an effective tool for poverty alleviation thereby contribute to the Millennium Development Goal of inclusive development. Further collaboration with the global partners is required to be continued for enhancing the quality and productivity of silk matching international standards.

The future of silk in the 21st century seems to be bright. Its consumption is likely to rise despite its high prices. However, there are doubts as to if the silk production will keep pace with the shooting demand. Japan is foremost in silk consumption. Besides, Chinese enjoying high standard of living are also boosting demand for silk within the country. In case of India, about 85 per cent of silk produced is consumed within the country and it has to import up to 5,000 MT every year to meet the requirement of its domestic industry demand. The ecological factor is pushing demand for silk in Europe and USA. The Middle East Arabs adore silk for the flowing robes, prayer mats and carpets. From this trend, it is clear that the global production may not increase much during the twenty-first century. As the market is likely to increase, it would provide and window of opportunity to countries having favourable factors for exploiting the potential for making the development inclusive.

Notes

1. FAO of the United Nations.
2. Final report of sub-group on Silk-National Fibre Policy, Ministry of Textiles, Government of India.
3. By Fiber Organon done in December 1998.

4

Sericulture: Indian Scenario

Historical Perspective

History of the Indian Sericulture industry dates back to the second century B.C. It is reported that raw silk was exported from India to Rome during the reign of Kanishka in B.C. 58. The Silk trade was mainly a caravan trade by land with two possible routes, one via Khotan across the Himalayas to Kashmir, Kandhar and Kabul and the other one via Kashgar and Yarkand, to Sogdiana and Bactria. The first one was the principal channel of the silk trade in the first century B.C. It appears that Kashmir[1] played an active part in silk trade and to some extent added to the commodities from India's stocks for export purposes. Ultimately, the silk passing through the above countries found its way to Syria where it was reworked for the Roman market. The chief traders between India and the Persian Gulf were Messenians. In the beginning of the Christian era, the silk trade was mostly undertaken through the Indian Ocean to the West due to popularity of sea routes. By then Alexandria became the central distributing mart of southern Europe and trade between India and Rome became the source of maritime prosperity with balance of trade in favour of India. The silk either in raw form or spun into thread, or woven into cloth, comprised a major part of the trade.

The Persians extended their territory and held monopoly in silk trade. The Romans had to depend upon them for all kinds of silk by the beginning of third century AD. India's role as an intermediate mart between China and Central Asia on the one hand, and Persia and Rome on the other was very significant. This continued till the sixth century A.D. In the seventh century A.D. strained relations between Persians and Greeks got normalized, which led to opening of Caspian Sea route from North China. In Eighth century, silk trade ground different route and Sindh was no longer intermediate transmitting country for Chinese silk. The ancient route from Central Asia through Kashmir and Kabul ceased to be the caravan route after ninth century.

Subsequently in the thirteenth century, after the Mohammedan conquest of Northern India, commercial relations between Kashmir and Khotan developed to a great extent and silk became one of the important commodities of export. Western countries continued to get their supplies of silk directly from China through the Caspian Sea. After conquering the greater part of Indian Peninsula in thirteenth century and the two following centuries, the Mohammedan Moors engrossed the entire commerce of the country and transmitted the rich merchandise of India including silk to the European markets, mainly from Aden in Arabia, by Red sea to Suez Canal, then by canals to the Nile, and there shipped for Grand Cairo and Alexendria. Silk was also transported from Persian Gulf to the Euphrates and there from by caravan to Aleppo and ultimately to the Western world. England was supplied with Indian commodities including silk by an annual ship from Venice, which was considered as a commodity of immense value and to be sold at their own prices.

As per the travel document of Ludovico Divarthema, large quantity of stuffs of cotton and silk was imported from a city called Bangechella in India major in the beginning of sixteenth century (1503). He also mentioned Combeia (presently Cambay) where from a large quantity of silk stuffs were exported to different parts of India and abroad. Thus, in the first phase, the Mohammedan Moors and Arabs were the chief oriental merchants and navigators. In the second phase, silk was exported to Mecca and other places from India. As a result of the discovery of the sea passage to India by the Portuguese in 1497, the mart was diverted from Venice to Lisbon. During fifteenth century, Calicut became the principal center for export of silk from India. The Portuguese established themselves there and for sometime became the principal distributors of Indian silk to Europe. Large quantities of silk were shipped from various Indian ports to Lisbon for distribution throughout thc Europc. By the end of sixteenth century, other European nations also joined trade struggle in India and several competitors took up trade of silk.

English commercial enterprise commenced in India in the beginning of the seventeenth century. In the first instance, the British took up export of silk and other commodities to England from Surat. At that time, Bengal was the center of Indian silk production with a fairly large silk industry. The first Letter Book of the East Indian Company shows that there was a favourable demand for Indian silk in England at that time and Merchants of the London Board insisted upon their agents in India for buying silk, which was an useful commodity. Chinese and Persian silk was also very popular in England. In 1616, Sir Thomos Roe, the first English Ambassador to the court of the Great Mughals wrote to their Directors about the silk of Bengal and need of a factory in Bengal. Silk produced in India was very cheap and used all over India. In 1620, Hughes, one of the directors of the East India Company came across the Bengal silk and suggested that variety of soil along with unlimited availability of cheap labour in Patna provides scope for producing silk for English uses. It was reported that best type of silk was produced from cocoons grown in Murshidabad of Bengal. The sale of the skein silk was a monopoly of the Governor and all transactions were carried through his agents. The Portuguese were also keen for dealing with silk. On account of the rising demand for silk by two rival nations, prices of silk was favourable for the Indian Merchants. By the middle of the seventeenth century, the East India Company started taking particular interest in Bengal silk. In the beginning, the company was interested in Tasar silk only. As part of their export trade, the company ordered for some quantities of Muga silk, which was introduced in 1624. Around 1646, the Dutch also gained prominence in silk trade with Persia. Due to decline in English silk trade with Persia, the East India Company paid more attention to Bengal silk and showed interest in production and manufacturing of this silk.

The silk trade of the company passed through a series of struggles till 1676. Streynshan Master, the chief representative of the company in Bay of Bengal, had referred to wrong practices by the servants of the company in buying up private supplies of silk and employing weavers to the detriment of the company's interest, which deteriorated the export trade of silk to a considerable extent. He stressed on the pricing of silk goods and proper supervision of raw silk. Numbers of regulations were framed with a view to increase silk trade of the company. England received large quantity of raw silk and silk goods from India and there was a gradual development of trade of silk between the two countries. Competition with the Dutch and other European nation resulted in the increase of the Indian silk trade of the company. This helped in promotion of production of raw silk in India.

Import of silk from India to England has been reflected in a pamphlet published in 1697 by T.S., a weaver in London, which had mentioned that it seemed that a large amount of money was invested in these commodities and a tremendous trade was carried on, which was supposed to be detrimental to the home manufacturers of England and therefore, there was an outflow of gold and coins to India for purchase of those commodities. The prosperity of silk trade did not last long and there was a set back to the trade due to rampant corruption by the agents and servants of the company engaged in the trade and also due to resultant price rise and quality deterioration of silk goods. To compensate the manufacturing interest of England and to supply them cheap materials, the Directors of the Company formulated a scheme, which tended to destroy or at least essentially impair the whole manufacturing interest of Bengal. Winders (silk reelers) were prevented from working in their private houses and were confined to the company's factories for company's benefit. It has been recorded[2] that in the 17th century silk weaving was an important cottage industry in Ahmedabad, Surat, Agra. Lahore and Fatehpur and Chinese silk was being used for weaving in all these places. There are reference to a flourishing silk weaving industry in Varanasi and delicate silk fabrics being exported[3] from Varansi to St. James and Versailles.

Silk trade experienced a set back during the French Revolution due to huge reduction in consumption of raw silk. The weavers suffered heavily and unemployment became common. Till 1833, Bengal silk industry saw some improvements as a result of entire cessation of Italian silks in England. There are references[3] to 11 head factories in the major silk producing areas under the commercial resident whose duty was to provide silk to company's filatures (2039 basins) and hired filature (3684 basins). Mulberry cocoon was brought through middlemen, called "pykars" to whom advances were being made. In 1835, the company's exclusive silk trade with India came to an end and raw silk trade was thrown open to private individuals. There were considerable fluctuations in silk trade till 1874 due to depression in European production and prevailing local conditions. However, there was a fairly steady increase in export of Indian silk to English market from 1866 to 1874. In 1875, exports declined and thereafter the Indian silk trade could not recover its former position. After succeeding in utilization of yarn from waste silk in 1877, export from India were confined to large quantities of waste silk. This led to decline in overall value owing to difference in prices. Japan and France took action for improvements in the silk production, which increased silk production considerably. However, Indian silk industry remained in a state of partial decay and demand for Indian silk practically disappeared from European market except for some special purposes.

Sericulture industry of Bengal was affected by an epidemic may be due to outbreak of "*Pebrine*" disease, which led to closure of silk production temporarily around the middle part of the 19th century. According to Prof. Lefroy's Report on Indian silk, Bengal silk industry lingered on in this condition till 1899, when it was decided to revive it. Thereafter, different measures were adopted to improve the position.

During the same period, the industry flourished in Kasmhir and Mysore. In Kashmir, univoltine silkworm races were reared and the industry grew under the royal patronage of Maharaja Ranbir Singh. The first ever sericulture department was started and silk reeling was taken up with imported machineries. Services of one British expert, Thomas Wardle, were availed for promoting silk industry in the Kashmir valley.

Credit for promoting silk industry in the south India goes to the far sighted policy of Tipu Sultan, who had sent persons from Tiruchirapalli to Bengal for learning the technique of silk production in the middle of 18th century. Later the British promoted mulberry cultivation in an organized manner in 1871 in the Madras Presidency. The District collectors were sanctioned fund for raising mulberry plantation. Selected rearers and silkworm breeding materials were brought from Bengal for taking up rearing. There were references to the efforts made[4] for introducing sericulture in the Madras presidency area.

Prior to independence, sericulture industry in India was limited to few pockets and was not organized. The level of technology was low. There are references to rearing of number of indigenous silkworm races, where productivity was low. Rearing of non-mulberry silk worms, like Tussah silkworm, Eri or Arindy, and Muga silkworm were popular in certain areas but were of local interest.

Modern Indian Silk Industry

India has the unique distinction of being the only country in the world bestowed by nature with all four varieties of silk namely, mulberry, tasar, eri and muga. The country is second next only to China in the production of raw silk. It has monopoly for production of the famed golden silk 'Muga', in the world. Of the four varieties of silk, the mulberry silk represents more than 90 per cent of the country's total raw silk production. A map showing sericulture activities in various states of India is given at Fig. 4.1.

Mulberry Silk

Mulberry silk is produced by silkworm called *Bombyx mori*. Depending on the mumber of generations in a year, it is termed as Uni-, Bi- or Multi- voltines. All the three voltines are grown in India. Mulberry silkworms feed on mulberry leaves and are monophagous. *Bombyx mori* is a completely domesticated worm and is to be fed and reared indoors. Mulberry sericulture dominates the field of sericulture in various aspects like quality of production, quantity and popularity. Therefore, often mulberry silk is called silk and normally sericulture refers to this mulberry culture and silkworm rearing. States of Karnataka, Andhra Pradesh, West Bengal, Tamil Nadu and Jammu and Kashmir accounted for more than 95 per cent of the country's total raw silk production.

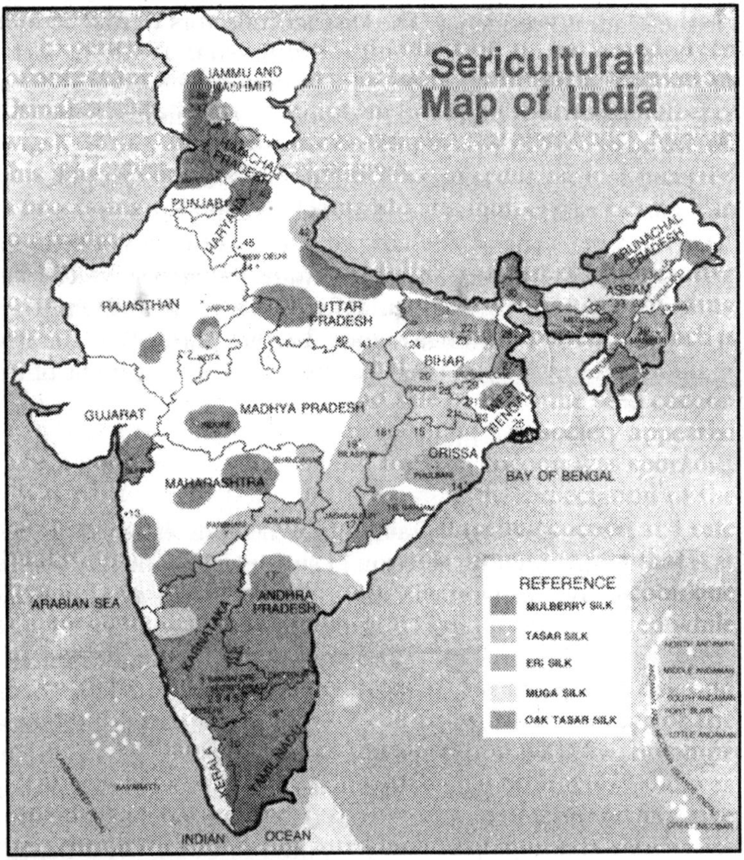

Source: Central Silk Board.

Fig. 4.1: Seri Map of India

Tasar Silk

Tasar Silk is produced by tasar silkworm (*Antheraea mylitta)* in the tropical area. The tasar silkworm feed on leaves of three primary food plants namely *Terminalia tomentosa* (Asan), *Terminalia arjuna* (Arjun) and *Shorea robusta* (Sal). Tasar is copperish coloured silk. The principal tasar producing States are Bihar, Madhya Pradesh and Odisha. Tasar culture is also practiced in Maharashtra, West Bengal and Andhra Pradesh on a smaller scale. In the temperate climate, the silkworm is called

Antheraea proylei. The Oak tasar culture is practiced in Sub-Himalayan States like Manipur, Himachal Pradesh, Uttar Pradesh, Assam, Meghalaya and Jammu and Kashmir. Tasar silkworm rearing is practiced by mostly by the families belonging to the Scheduled Tribes as a socio-cultural practice even prior to the introduction of mulberry sericulture in India. It was practiced as a part of rich traditional culture. Unlike mulberry sericulture, where 60 per cent of the cost of the cocoons accounted for growing of food plants, in tasar culture 14 million hectares of plantation available in the natural forest in tropical and temperate zones are used for silkworm rearing. In some parts of the country, plantation of tasar host plant has been raised for benefit of the tasar rearers.

Eri Silk

The Eri Silk is also known as Endi or Errandi. It ranks next to tasar in commercial production. The Eri silkworm (*Phiosamia ricini*) feeds on castor leaves raised in area up to an altitude of 5000 ft. above sea level. It is multivoltine in nature and 4 to 6 crops are grown annually. The silkworm is domesticated and rearing is done indoors. Its wild counterpart is found in North-Eastern India. Besides Kesseru, Tapicoa, Cassava, Papaya, Papyam and Barkessuru are also sometime used. Eri cocoons are white and brick red in colour with shades, varying according to the fauna and plant used. Bulk of eri silk is produced in Assam, while on small scale it is also produced in Bihar, West Bengal, Odisha and Manipur.

Muga Silk

Muga is obtained from the worms (*Antherea assamensis*), which are similar to *A. mylitta* and *A. proylei*. This silkworm feed on the leaves of Som (*Machilus bombycina* King) and Sualu (*Litsea polyanthia* King), which are widely distributed in the valley of river Brahmaputra. The muga silkworm is multivoltine in nature and 5 to 6 crops are grown per annum. The silkworm is semi domesticated and the practice is exclusive to Assam. The cocoons are fawn coloured and give beautiful golden yellow colour yarn on reeling.

Sericulture Development Under the Five Year Plans

In view of its multifarious advantages, promotion of sericulture was given attention after independence. The Central Silk Board was set-up by the Government of India in the year 1949 with a view to promote sericulture activities in general and provide required technical help by undertaking Research and development activities in particular.

First Five Year Plan

In the First Five Year Plan, promotion of sericulture was dealt under 'Other Village Industries'. Central assistance was provided to the states, both as grants and loans, with a view to consolidate the industry and set up required organizations. Production of raw silk increased by about 66 per cent from 8.94 lakh kg to 14.86 lakh kg during the First Five Year Plan period.

Second Five Year Plan

Development of infrastructure for production of quality silk worm egg and reeling of cocoon were given stress in the Second Five Year Plan. A special legislation was enacted in Karnataka State to regulate quality of silk worm seed and to prevent the use of unexamined seed for commercial production. Besides, stress was laid on improvement of cottage basins for improving the quality and output of raw silk. All India Sericulture Training Institute was set-up under the Central Silk Board at Mysore. An import of raw silk was channelized and its distribution was regulated to stabilize price of silk yarn in the domestic markets. A scheme was introduced to promote export of silk fabrics under the Central Silk Board. Production of raw silk increased to 15.13 lakh kg during the plan period.

Third Five Year Plan

Keeping in view the importance of sericulture, it was assigned a separate place in the Third Five Year Plan and a separate head of account was given under the budget head of "Village and Small Scale Industries" during the period. Strengthening of seed organization and promotion of export were given importance in this plan. For development of manpower and technical know-how, the Central Sericulture Research and Training Institute at Berhampore and the Basic Tasar Seed Station were set-up. Quality control, inspection and standards were introduced in 1965 for boosting up export. Output of raw silk increased to 20.65 lakh kg by the end of the third plan period.

During the transitional period (1966-67 to 1968-69), a scheme for supporting price of tasar cocoon was introduced for giving a fair price to the tasar growers, most of whom belong to Scheduled tribe and to sustain the industry. Output of raw silk has increased to 23.20 lakh kg by end of 1968-69.

Fourth Five Year Plan

Increase productivity, reduction in cost of production, rationalization of production techniques and creation of additional employment opportunities were taken as the key areas in the Fourth Five Year Plan. Attainment of self-sufficiency and meeting the domestic demand for raw silk were given specific attention. Arrangements for production of seed and research were further strengthened. Attempts were made to improve quality of silk by promoting rearing of bivoltine breed of silkworm in place of multivoltine breed in the traditional silk producing areas of Karnataka, Tamil Nadu and West Bengal. Production of raw silk went up to 28.94 lakh kg and earning from export reached Rs. 14.46 crores by the end of the Fourth Plan period.

Fifth Five Year Plan

Keeping in view the multifarious advantages of mulberry sericulture, special attempts were made in the Fifth Five Year Plan for introducing mulberry sericulture in new and non-traditional areas. Under a central programme, initiated in 1975-76, arrangements were made for supplying mulberry cuttings or saplings to the rearers. Besides research and extension programme on improvement of quality by introduction of bivoltine pure and cross breed races was also intensified. Central Silk Board established 11 silk worm grainages for production of silkworm eggs for supply to the rearers. Silk yarn price stabilization scheme was introduced and operated by the CSB in 1977-78. This scheme envisaged provision of a fair price to the primary producers and supply of raw silk at a fair and stabilized price to the silk handloom weavers. Production of raw silk increased to 37.11 lakh kg and export earnings increased to Rs. 33.06 crores, by the end of the Fifth Plan. The Central Silk Board implemented an Intensive Sericulture Development Project during 1978-80 as a result of which production of raw silk rose to 48.05 lakh kg and export of silk went up to Rs. 48.83 crores.

Sixth Five Year Plan

Emphasis on improvement of quality of mulberry silk continued during the Sixth Five Year Plan (1980-81 to 1984-85). With a view to undertake research and evolve technology suiting the need of tropical areas International Centre for Training and Research in Tropical Sericulture (ICTRETS) was set-up in 1980 at Mysore under Indo-Swiss Technical and Scientific Co-operation Programme. Besides, The National Silkworm Seed Project was started in 1981 to expand seed organization and production of quality bivoltine silkworm seed. A number of seed production centers and basic seed farms at P3 and P2 levels were set up for maintaining vigour and quality of seed production in different parts of the country. Central Silk Technological Research Institute was set-up at Bangalore in 1982-83 to evolve suitable technology in the post-cocoon (reeling, twisting etc) stage. Production of raw silk went up to 76.73 lakh kg and silk export earnings to Rs. 129.05 crore by the end of the Sixth Plan period.

Seventh Five Year Plan

By the end of the Sixth Plan (1984-85), 6,754 MT of raw silk was produced against target of 9,000 MT and export of silk went up to Rs. 129.05 crores against target of Rs. 100 crores, with the generation of 1.32 million man days agricultural and industrial activities of sericulture during the Sixth Plan. However, production of better quality Bivoltine silk, which was initiated over a decade back, with the objective of replacing low yielding and poor quality of multivoltine races could not make any break through. Despite these efforts, bivoltine production would only reach a level of 170 MT by 1984-85 against a target of 1000 tonnes. Besides, progress under the World Bank Project in Karnataka undertaken during the sixth plan was hampered due to the non-availability of high grade silk-worm races, lack of good quality seed cocoons.

Promotion of sericulture during the Seventh Five Year Plan stressed on creation of infrastructure support particularly for production of basic quality seeds, taking up mulberry plantation in irrigated land, provide better extension sericulture facility with stress on bivoltine silk production. A target of production of 9,980 MT mulberry and 920 MT non-mulberry raw silk was kept for the Seventh Plan. The National Silkworm Seed Project (NSSP) of the Central Silk Board was strengthened further to play a greater role and meet at least 5 to 10 per cent of the total seed requirement of the industry. Greater attention was paid to develop post-cocoon technology so as to improve the quality of silk and fabric produced.

National Sericulture Project

With a view to accelerate the production of quality mulberry silk in the country, a World Bank/Swiss assisted National Sericulture Project was implemented at a total cost of Rs. 555.30 crores over a period of 7 years from 1989-90. Out of this, the provision under Government Budget was Rs. 389.70 crores and the balance Rs. 165.70 crores was Institutional Credit for both "On farm" and "Non-farm" activities. The National Sericulture Project had 6 components, of which one component was implemented by the Central silk Board under the Central Sector in 5 traditional states and 12 Pilot States and the other five state components by the State government of Karnataka, Andhra Pradesh, Tamil Nadu, West Bengal and Jammu and Kashmir.

The Project, aimed at strengthening of infrastructure for research, extension, seed production, marketing of cocoons and raw silk processing for bringing about an improvement in quality and productivity of Indian silk. Besides, the project envisaged-extension of financial support to rearers, reelers, twisters and seed producers through financial institutions. The project also aimed at encouraging involvement of women and NGOs in sericulture development. National Sericulture Project envisaged–

(i) Development of 5,76,000 hectares of additional mulberry plantation, (ii) Increase production of cocoons to 58,700 tonnes, (iii) Additional production of 6,000 tonnes of raw silk, (iv) Additional employment of 1 million persons, (v) Increase in Export earnings to Rs. 900 crores in quantitative terms. It also envisaged– (i) Improvement in quality and productivity of Indian silk, (ii) Strengthening of infrastructure for research, extension, seed production, marketing of cocoons and raw silk processing and quality control, (iii) Providing financial support to rearers, reelers, twisters and seed producers through support of NABARD and IDBI, (iv) Strengthening socially desirable features such as employment of women, involvement of Non-Governmental Organizations (NGOs) in sericulture development, etc.

Apart of promotion of sericulture in the traditional States, the project also aimed to introduce sericulture and demonstrate technologies potentialities for future development or expansion of sericulture industry in 12 new and non-traditional (Pilot) states namely Maharashtra, Kerala, Odisha, Assam, Bihar, Uttar Pradesh, Madhya Pradesh, Rajasthan, Gujarat, Himachal Pradesh, Punjab and Haryana.

Eighth Five Year Plan

By 1991-92, production of mulberry silk went up to 12,450 tonnes, which made India the second largest silk producer of the world. The sericulture was practiced in 52,000 villages on 3.32 lakh hectares of irrigated and rainfed land under mulberry employing about 55 lakh persons in planting, seed preparation, silkworm rearing, reeling, weaving and other operations. Mulberry silk, accounting for more than 90 per cent of the total silk produced in the country, is concentrated in the States of Karnataka, Andhra Pradesh, Tamil Nadu and West Bengal. However, imports of silk doubled from about 1181 tonnes in 1984-85 to about 2,500 tonnes in 1989-90. A National Sericulture Project costing Rs. 555 crores with assistance from IBRD/IDA and Swiss Development Cooperation was taken up in 1989-90.

A target for production of 21,400 tonnes raw silk, including 2,000 tonnes of bivoltine variety, was fixed for the Eighth Plan. Under the earlier plan, productivity of mulberry silk increased from 32.09 kg/ha in 1984-95 to 38.29 in 1989-90. The plan envisaged to increase productivity further to the levels of sericulturally advanced countries like China and Japan. Production of bivoltine cocoons was particularly stressed for meeting the seed requirements, larger production of quality silk and lessening dependence on imports.

Besides, in order to prevent spread of pebrine disease and contain it, emphasis was proposed to be laid on thorough moth examination and providing assistance to sericulturists and private graineurs. Post cocoon technology like reeling of cocoon was given greater attention. Provision of working capital for the purchase of cocoons was given stress for meeting the need of reelers. Further, in order to remedy the regional imbalance and to improve the availability of raw silk to weavers, schemes for silk production would be taken up in new areas in States like U.P., Bihar, Assam and Odisha in addition to traditional silk growing states like Karnataka, Andhra Pradesh and Tamil Nadu.

As a result of schemes implemented during the Eighth Five Year plan, production of mulberry silk production went up to 12,954 MTs, Tasar silk to 235 MTs, Eri silk to 864 MTs and Muga to 73 MTs, total being 14,126 MTs by the end of the plan period in 1996-97.

Ninth Five Year Plan

Production of raw silk in India went up to 14,000 MT in 1996-97, by the end of the Eighth plan. Mulberry accounted for 92.5 per cent, eri 5.3 per cent, tasar 1.7 per cent and muga 0.5 per cent of the total raw silk production in the country. Silk worth Rs. 880.44 crores were exported in 1996-97. Sericulture sector provided employment to 61 lakhs persons in 1996-97.

A target for production of 16,350 MT and 20,540 MT of raw silk was fixed for the first year (1997-98) and the terminal year (2001-02) of the Ninth Plan respectively. Target for employment in sericulture sector was kept at 62 lakhs and 71 lakhs for the first year (1997-98) and the terminal year (2001-02) respectively and same for export was kept at Rs. 1050 crores and Rs. 1525 crores respectively.

A scheme for establishing decentralized peripatetic sericulture centers for farmers was formulated by the CSB, under which farmers/rearers were trained in improved sericulture practices. With financial participation of the CSB and the Department of Rural Development (Ministry of Rural Areas and Employment), 13 Sericulture Training Schools for farmers were set-up in non-traditional sericulture States with a view to provide required advice, technical guidance and knowledge about sericulture.

The Central Silk Board formulated 36 catalytic development schemes for implementation in the Ninth plan to motivate the States to increase productivity and quality and provide marketing support. These schemes aimed at increasing quality of raw silk production, promotion of drip irrigation, enterprise development, improved variety of mulberry etc. Madhya Pradesh and Manipur Government have implemented sericulture projects with help of the Japan government (JBIC) at a cost of Rs. 748.8 and Rs. 490.61 crores respectively. Under the bivoltine sericulture development project taken up with assistance from Japan International Cooperation Agency (JICA), a number of improved bivoltine breeds were evolved of which bivoltine hybrids produced with the races of viz., CSR 2, 4, 5, CSR 12, 6, CSR 18,19, CSR 16, 17, CSR 3, 6 were released for commercial use. These hybrids yielded better quality (2/3 A grade) silk during experimentation.

The Catalytic Development Programme (CDP) was revamped by– (i) simplification of the schemes, (ii) widening the scope and scale of assistance, especially for the bivoltine programme, (iii) weeding-out schemes with low levels of acceptability with state governments, (iv) disbanding to a large extent, capital loan assistance—instead, integrating use of financial assistance available for entrepreneurs from State Financial Corporations/Financial Institutions/Commercial Banks for capital investment, and (v) rationalizing the varied proportions in sharing subsidy by CSB and States. Rs. 49.95 crores were spent during the plan against Rs. 58.84 crores allocated for the plan.

Against a target for production of 20,666 MT of raw silk during the Ninth Plan, 17,980 MT of raw silk (87% of the target) was produced in the terminal year of the plan (2001-02). Details on year wise production of different type of silk are given in Table 4.1.

Table 4.1: Production of raw silk

(in MT)

Silks	Target	1997-98	1998-99	1999-00	2000-01	2001-02
Mulberry	18954	14048	14260	13944	14432	16400
Tasar	335	312	242	211	237	335
Eri	1264	814	970	974	1089	1150
Muga	113	62	72	85	99	95
Total	**20666**	**15236**	**15544**	**15214**	**15857**	**17980**

Source: Central Silk Board.

So far as productivity of mulberry silk is concerned, it has increased steadily from 451.7 kg of cocoon and 49.7 kg of raw silk per ha at the beginning of the plan to 601.59 kg cocoon and 68.26 kg of raw silk per hactre of plantation respectively by (2001-02) the end of the 9th plan. Details on productivity given in the Table 4.2.

Table 4.2: Productivity of Mulberry Sericulture

Year	Area under Mulberry (in ha)	Mulberry cocoon produced (in MT)	Mulberry raw silk (in MT)	Non-Mulberry raw silk (in MT)	Total raw silk (in MT)	Productivity of silk cocoon (in kg/ha)	Productivity of raw silk (in kg/ha)
1997-98	282,244	127,495	14,048	1,188	15,236	451.71	49.77
1998-99	270,069	126,566	14,260	1,284	15,544	468.64	52.80
1999-2000	227,151	124,531	13,944	1,270	15,214	548.23	61.38
2000-2001	215,921	124,663	14,432	1,425	15,857	577.35	66.84
2001-02	232,076	139,616	16,400	1,580	17,980	601.59	68.26

Source: Central Silk Board.

Sericulture Development Under the New Textile Policy 2000

The New Textile Policy of the Government of India, launched in 2000 seeks to focus on the following for promotion of the sericulture industry in general and achieving international standard in all varieties of silk in particular.

Tenth Five Year Plan

Importance of sericulture as a labour-intensive, agro-based industry providing employment to about 6.25 million persons has been acknowledged during the preparation of the Tenth Plan. Difficulties in achieving the target fixed for production of raw silk during the Ninth Plan were attributed to shortfall in the demand for multivoltine silk on account of the handloom and power loom weavers switching over to using imported Chinese silk for weft and warp purpose.

Acknowledging the lower production cost of raw silk in China as a major threat to the Indian silk industry and growing international demand for silk as the major opportunity, strategy[6] for development of the sericulture industry during the Tenth Plan envisages several specific initiatives namely– (i) achieving international standards in all varieties of silk, (ii) improving the Research and development and effective transfer of technology, (iii) focus upon increasing bivoltine silk production and non-mulberry varieties of silk, and (iv) encouraging cluster development for reeling and weaving and strengthening linkages between sericulture and textile industry.

Approach paper to the Tenth Plan underlined the need to promote the culture of quality in every production process and take up market linked production planning. China is the major producer of silk and accounts for 75 per cent of global production. It dominates export of raw silk, silk yarn and silk fabrics with a share of about 93 per cent, 43 per cent and 29 per cent of the world trade respectively. The Indian sericulture industry continued to face major problem due to import of silk from China through legal as well as illegal ways, which is affecting price of raw silk in the country and return to farmers/sericulturists.

Keeping in view the emerging trends in the world of silk industry, improvement of quality and productivity has been assigned importance with particular stress on production of bivoltine silk.

Eleventh Five Year Plan

The Sub-Group on Sericulture recommended a target for production of 26,000 MT silk including 23,000 mulberry silk of which, 5000 M.T. to be bivoltine silk by 2011-12 the end of the Eleventh Plan. As against this, details of the achievement are given in the Table 4.3.

Allocation and Expenditure of plan fund on the Schemes and Programmes implemented by the CSB either directly or through the State Government during the Eleventh Plan are given in the Table 4.4.

From the expenditure pattern, it appears that whereas, the fund earmarked for the development of sericulture have been utilized in full, the physical achievements particularly quality and quantity of silk yarn production.

Silk yarn produced continues to remain suboptimal. The silk weaving industry has to depend on import particularly from China for meeting the requirement.

A brief account of the sericulture statistics of India is given in Table 4.5.

Quantity of mulberry and non-mulberry silk produced (year-wise) in the country since 1980s is given in Table 4.6.

From commercial point, mulberry silk continues to be the most important type of silk produced in India. An account of the key parameters on production of mulberry silk are given in Table 4.7.

Production of different category of silk varied from state to state on account of several factors. From available data, it is clear that mulberry silk continue to form bulk of the domestic silk production. Tasar silkworm is reared under wild conditions in the natural/manmade forest by rearers belonging to the Scheduled Tribes. Production of Tasar has been inconsistent due to a number of factors, such as denudation of forest, reduction in host plant, uncertainty due to outdoor rearing, low economic return, low level of technology and difficulty in attracting new rearers for taking up rearing etc. Production of the Muga and Eri silk remained somewhat static. However, mulberry sericulture has seen a steady growth in India.

Table 4.3: Achievement of Sericulture Sector

Particulars	*2007-08*	*2008-09*	*2009-10*	*2010-11*	*2011-12*
Mulberry Plantation Area (Lakh Ha)	1.85	1.78	1.84	1.7	1.81
Mulberry					
Bivoltine (mt)	1,175	1,250	1,200	1,400	1,685
Crossbreed (mt)	15,070	14,360	15,122	14,960	16,587
Total (Mulberry) (mt)	**16,245**	**15,610**	**16,322**	**16,360**	**18,272**
Vanya Silk					
Tasar (mt)	428	603	803	1,166	1,590
Eri (mt)	1,530	2038	2,460	2,760	3,072
Muga (mt)	117	119	105	124	126
Total (Vanya) (mt)	**2,075**	**2760**	**3,368**	**4050**	**4,788**
Total raw silk production (mulb+vanya)	**18,320**	**18,370**	**19,690**	**20,410**	**23,060**
Employment Generation (lakh persons)	61.2	63.1	68.17	72.5	75.6
Exports (in Cr. Rs.)	2767.87	3178.19	2892.44	2863.76	2285.08

Source: Central Silk Board.

Table 4.4: Allocation and Expenditure Incurred by CSB

(amount in Rs. in crores)

No.	*Programmes of CSB*	*XI Plan Allocation 2007 to 2012*	*XI Plan Expenditure 2007 to 2012*
1.	Research, Development, Training and IT Initiatives	144.55	144.60
2.	Seed Organisation/HRD	78.36	78.31
3.	Quality Certification System	14.75	14.75
4.	Catalytic Development Programme (CDP)	821.74	821.74
	Grand Total	**1059.40**	**1059.40**

Indian Position in Global Sericulture

Indian breeds/hybrids have the potential to produce the same quality, our system of sericulture practices is entirely different from those of the Chinese. The strict maintenance of discipline and better linkage from farmers to weavers, large-scale operation of egg production, reeling and weaving using modern machineries, strict control measures for diseases, uniform adoption of new technologies, supply of required quantity of quality eggs in time to avoid the chances of contamination of young silkworm etc. make the Chinese sericulture more vibrant, economically sound and sustainable. Another area of difference is that the entire production is state controlled with no open marketing/auction systems for silk cocoons and yarns. The comparative statistics (Table 4.8) indicate the strength of China in the areas of seed production and distribution system, higher unit area productivity and quality thus reducing the costs of production.

Silk yarn and fabrics have remained important components of international trade and commerce. Globalization and liberalization of trade had its impact on the silk trade as well. An account of export and import of silk yarn and fabrics are given in Tables 4.9 and 4.10.

Table 4.5: Sericultural Statistics in India

Particulars	*Unit*	*2008-09*	*2009-10*	*2010-11*	*2011-12(P)*
Mulberry Plantation	ha.	177943	183773	170314	182905
Production of					
Reeling Cocoon of which	MT	124838	131661	130714	–
Bivoltine	MT	8429	8080	9290	–
Cross Breed	MT	116409	123581	121424	–
Production of Mulberry Raw Silk	MT	15610	16322	16360	18186
Bivoltine	MT	1250	1200	1400	1796
Multivoltine	MT	14360	15122	14960	16390
Silk waste	MT	3746	4080	4090	4546
Productivity Parameters					
Reeling cocoon/hect	kg/ha.	702	716.43	767.49	–
Reeling cocoon/100 dfls	Kg/100 dfls	51.78	52.85	55.23	–
Rawsilk/hect	kg/ha.	87.73	88.82	90.2	–
Renditta	No.	8	8	7.99	–
Vanya Silk: Production of					
Tasar Reeling Cocoon	Lakh Kahan	5.33	7.45	10.29	–
Tasar Rawsilk	MT	603	803	1166	1577
Eri Cut Cocoon	MT	2593	2801	3526.67	–
Eri Spun Yarn	MT	2038	2460	2760	3110
Muga Reeling cocoon	Lakh Nos	5879.77	5155.29	6629.42	–
Muga Rawsilk	MT	119	105	124	127
Vanya Silkwaste	MT	800	980	1012	1203
Total Production (*Vanya* Silk)	MT	2760	3368	4050	4814
Total Rawsilk Production	MT	18370	19690	20410	23000
Production of					
Spun silk yarn	MT	500	560	585	–

(*Contd.*)

Table 4. 5 (*Contd.*)

Particulars	*Unit*	*2008-09*	*2009-10*	*2010-11*	*2011-12(P)*
Noil Yarn	MT	250	280	295	–
Production of Silk fabrics					
Qty	Lakh sq m.	4439	4477.28	4322.73	–
Value	Crore Rs.	11098	13431.84	15129.57	–
Silk Export					
	Crore Rs.	3166.45	2860.05	2885.82	2285.08
Mn. US$		692.88	603.13	633.62	478.90
Silk Import					
	Crore Rs.	1714.48	1810.50	1714.11	1629.64
Mn. US$		374.86	381.57	375.88	340.26
Employment	Lakh Persons	63.10	68.17	72.50	–
Infrastructure					
Number of Reeling units					
Filature/Cottagebasins	Nos.	26631	26631	26631	26631
Charka	Nos.	28014	28014	28014	28014
Multiend	Nos.	201	201	201	201
Number of Sericulture Villages	Nos.	50918	50737	52360	52360
Number of Sericulture families	Nos.	817605	925288	947631	947631
Prices					
Mulberry:					
Reeling Cocoon	/kg	135	171	233	224
Rawsilk	/kg	1285	1580	2125	1991
Silkwaste (Filature)	/kg	130-200	190-230	220-380	220-380
Vanya:					
Tasar					
Reeling cocoon (Rs./'000 Nos)	/000 nos	1000-1400	650-1050	1050-1150	1250-1500
Rawsilk	/kg.+	1600-1800	1550-1600	2200-2400	2000-2700
Eri					
Cut cocoon	/kg **	200-500	220-500	220-500	240-600
Rawsilk (spun yarn)	/kg ***	800-1400	1000-1200	1200-1300	1200-1300

(*Contd.*)

Table 4. 5 (*Contd.*)

Particulars	Unit	2008-09	2009-10	2010-11	2011-12(P)
Muga					
Reeling cocoon (Rs./'000 Nos)	/000nos	550-1200	800-900	700-900	1200-1500
Rawsilk	/kg	3000-6200	6000-8000	6500-7000	8000-9000

Note: ha–hectares, MT–Metric tonnes, kg.–kilograms, Nos.–Numbers,
P: Provisional
.+ Reeled Yarn
: Cut cocoon * Spun Silk

Source: CSB & DGCIS.

Table 4.6: Production of Different Types of Raw Silk

(In MT)

Years	Mulberry	Tasar	Eri	Muga	Total
1980-81	4593	265	135	48	5041
1990-91	11486	380	624	70	12560
2000-01	14432	237	1089	99	15857
2001-02	15842	249	1160	100	17351
2002-03	14617	284	1316	102	16319
2003-04	13970	315	1352	105	15742
2004-05	14620	322	1448	110	16500
2005-06	15445	308	1442	110	17305
2006-07	16525	350	1485	115	18475
2007-08	16245	428	1530	117	18320
2008-09	15610	603	2038	119	18370
2009-10	16322	803	2460	105	19690
2010-11	16360	1166	2760	124	20410
2011-12	18272		4788		23060

Source: Central Silk Board.

Even though there has been growth in export and import, quality of Indian silk continued to be an area of concern. India continued to import sizable quantity of raw silk of bivoltine variety for meeting the yarn requirement of silk textile industry, particularly for use as warp in the silk power loom. Besides, silk fabrics are also imported for meeting the need of the fashion and apparel industry.

From the data, it is clear that the import of raw silk and fabrics have gone up substantially. Quality and productivity of Indian silk need to be increased to level at par with those of China for checking this trend.

Table 4.7: Mulberry Silk Production in India

Years	Mulberry Plantation ('000 hectare)	Reeling Cocoons (in '000 MT)	Raw silk (in MT)	Silk Waste (in MT)	Renditta
1980-81	170	58	4,593	1,376	12.70
1990-91	317	117	11,486	3,953	10.20
2000-01	216	125	14,432	4,237	8.66
2001-02	232	140	15,842	4,655	8.81
2002-03	194	128	14,617	4,514	8.77
2003-04	185	117	13,970	3,764	8.41
2004-05	172	120	14,620	3,587	8.21
2005-06	179	126	15,445	3,749	8.17
2006-07	192	135	16,525	4,055	8.20
2007-08	185	135	16,245	3,416	8.12
2008-09	178	125	15,610	3,746	8.00
2009-10	184	132	16,322	4,080	8.07
2010-11	170	131	16,360	4,090	7.99
2011-12	183	–	18,272	4,788	–

Source: Central Silk Board.

Impact of Liberalisation

Quantitative Restriction (QR) on import/export of silk were removed with effect from 1st April 2001. There has not been any significant fluctuation in the prices of cocoons or yarn produced in the Country. However, it is rather too early to forecast as to how the market would behave in the next few years. Nevertheless, it is safe to assume that the future growth of the silk industry and export of silk goods would, to a certain extent, depend (in spite of our strong domestic demand for handloom silks) on how it reacts in the short, medium, and long-term to the market changes that are sure to affect its competitiveness in the post GATT regime. Production of Indian silk shall have to become more technology intensive and yet preserve its comparative advantage with respect to ethnicity, design and colour. Improvement in product quality and product diversification continues to be of primary importance. Obsolete production facilities are required to be replaced by modern equipment. Indian power loom silk weaving and processing sector has a 'machine vintage' problem with most of the machinery clearly older than 20 years.

Table 4.8: Comparative Mulberry Sericulture Statistics—China *vis-à-vis* India

Parameter	*China*	*India*
Area under Mulberry (lakh ha.)	7.36	1.79
Leaf yield (MTs/ha/yr)	25-30	30-50
Races reared	All bivoltine	Mostly cross breeds
Egg production (crore dfls)	75.962	25.65
Supply system	Majority chawki reared	Majority supplied as eggs
Time of supply	Batch-wise	Throughout the year
No. of crops per year	2-3: Temperate 6-8: Tropical	5-6 8-12 (Split plot system)
Dfls brushed/ha/yr	1050	1492
Cocoon yield (kg/100 dfls) (2 boxes)	75.92	51.0
Leaf cocoon raio (kg)	16-18	20-22
Cocoon yield (kg/ha)	736.89	698.0
Cocoon weight (g)	1.9-2.0	1.6-1.8
Cocoon shell percentage (%)	21-23	CB: 17-19 BV: 20-22
Filament length (m)	>1000	<1000
Renditta	6.4	8.2 (6.9 for bivoltine)
Grade of silk	A-4A	Un-graded to A grade
Cost of cocoon production (Rs./kg)	40-45	65-70
Prevailing cocoon price (Rs.)	85-90	90-120
Raw silk production (kg/ha/yr)	114.87	85.02
Av. Crop loss/year (%)	3-5	10-15
Demand and supply position	90% Export	Self consumption 30 % deficit
Raw silk produced (MTs/yr)	86500	15445

Source: Li Long and Hu Zhuozhong, Indian Silk, May 2006.

Table 4.9: Item-wise Export of Silk (2007-08 to 2011-12)

Item	*Rs Crore*					*US $ Million*				
	2007-08	*2008-09*	*2009-10*	*2010-11*	*2011-12 (P)*	*2007-08*	*2008-09*	*2009-10*	*2010-11*	*2011-12 (P)*
Natural Silk Yarn, Fabrics and Made-up	1540.93	1664.82	1411.12	1660.8	949.02	382.56	365.08	297.76	364.68	198.74
	*	8.00%	-15.2%	17.70%	-42.9%		-4.6%	-18.4%	22.50%	-45.5%
Readymade Garments	1093.67	1437.73	1383.42	1169.26	1267.08	271.97	314.3	291.48	256.69	266
	*	31.50%	-3.8%	-15.5%	8.4%		15.60%	-7.3%	-11.9%	3.60%
Silk Carpet	72.11	58.67	40.59	11.69	19.21	17.97	12.38	8.55	2.58	3.99
	*	-18.6%	-30.8%	-71.2%	64.30%		-31.1%	-30.9%	-69.8%	54.70%
Silk Waste	12.15	5.23	24.92	44.07	49.77	3	1.12	5.34	9.67	10.17
	*	-57.0%	376.50%	76.80%	12.90%		-62.7%	376.80%	81.10%	5.20%
Total	**2718.86**	**3166.45**	**2860.05**	**2885.82**	**2285.08**	**675.5**	**692.88**	**603.13**	**633.62**	**478.9**
	*****	**16.50%**	**-9.7%**	**0.90%**	**-20.8%**		**2.60%**	**-13.0%**	**5.10%**	**-24.4%**

* Figures given in the row indicate the percentage change over the previous year.

Table 4.10: Item-wise Import of Silk (2007-08 to 2011-12)

Item	*Rs Crore*					*US $ Million*				
	2007-08	2008-09	2009-10	2010-11	2011-12 (P)	2007-08	2008-09	2009-10	2010-11	2011-12 (P)
Raw Silk	734.44	901.74	933.7	927.59	1111.98	182.3	197.03	197.1	203.57	232.22
	*	22.80%	3.50%	-0.7%	19.90%		8.10%	0.00%	3.30%	14.10%
Silk yarn and fabrics	833.07	812.74	876.8	786.52	517.66	206.89	177.83	184.47	172.31	108.04
	*	-2.4%	7.90%	-10.3%	-34.2%		-14.0%	3.70%	-6.6%	-37.3%
Total	1567.51	1714.48	1810.5	1714.11	1629.64	389.19	374.86	381.57	375.88	340.26
	*	9.40%	5.60%	-5.3%	-4.9%		-3.7%	1.80%	-1.5%	-9.5%

* Figures given in the row indicate the percentage change over the previous year.

It can also be said that over the years, as a result of strong and tradition-bound home demand, Indian silk industry has been to a large extent insulated against international competition and cushioned against low quality. In a way, it has had a fairy-tale run for a fairly long time. With the removal of Quantitative Restrictions, the question has come up if Indian silk can compete in its home base, let alone in the international market. Like for any other product, the mantra for the future is going to be higher quality at competitive prices.

Entry of China in to the WTO has brought in new opportunities as well as challenges to the Indian silk industry. It would help India to improve its global positioning. It is generally believed that the Chinese silk is highly subsidized and its products are sold at cheaper rates compared to Indian silk. Under the Subsidies and Countervailing Measures of the WTO agreement, China would have to regulate these subsidies. Nevertheless, it is beyond doubt that both productivity and quality of Chinese silk is much higher than the Indian Silk. This has added new dimensions to the target of the Tenth Plan underlining the urgency to adopt bivoltine sericulture on a large scale and improve the quality of raw silk.

Anti-Dumping Duty

Raw Silk

During the year 2001-02, there was large scale dumping of Chinese Raw silk into the country de-stabilized the Indian silk Industry. In order to safeguard the interest of the domestic silk industry against the cheap imports, a petition was filed by CSB with Directorate General of Anti-dumping and Allied Duties (DGAD). The petition was filed on behalf of various silk reelers Associations along with the State Governments of Karnataka, Andhra Pardesh and Tamil Nadu for imposition of anti-dumping duty on raw silk imported from China P.R. Considering the merits of the case, the antidumping duty was imposed during January, 2003 (*for a period of 5 years*) on raw silk of 2A grade and below imported from China on the cheap imports with a reference price of US$27.27/kg. The anti-dumping duty so imposed was in force until Jan-2008.

Subsequently, a sunset review has been taken up for continuation of anti-dumping duty on import of raw silk. Accordingly, anti-dumping duty has been further continued with an enhanced reference price of US$ 37.32 per kg. and the same will be effective till January-2014.

Silk Fabrics

During the year 2002-03, it has also been brought to the notice of the Central Silk Board that a large scale dumping of silk fabrics was taken place from China PR creating a situation wherein, large number of powerloom silk weavers of Karnataka, Uttar Pradesh and Gujarat were displaced since they had to shut their looms. These goods were being dumped by the Chinese exporters.

Central Silk Board in associations with the various silk Powerloom Weavers' Associations situated in Karnataka, Gujarat and Uttar Pradesh field a petition with the Director General of Anti-dumping and Allied Duties (DGAD), Ministry of Commerce (GOI) during September 2004.

After investigations the DGAD, vide GOI Gazette Notification No. 256. dated 16th November-2006 has imposed anti-dumping duty on silk fabrics (20-100 gms) imported from China P.R. The final duty has been imposed taking into consideration the dumping margin range from 42 per cent to 77 per cent. The reference price-range on the fabric weight ranging from 20 to 100 gms. per meter indicating a reference price ranging between US$1.662 and US$ 4.526 per meter. Any silk fabric imported from China R P below these reference prices attracted Anti-dumping Duty. The duty so imposed was in force till Dec-2011.

Further, based on the request of the CSB and the domestic powerloom silk weaving industry a sunset review application for continued imposition of anti-dumping duty for another term of 5 years was filed. Considering this, The Directorate General of Anti-dumping and Allied Duties (DGAD) had extended the imposition of anti-dumping duty for another term of 5 years till December 2016.

Industry Trends[7]

During 15 years spreading over 1996 to 2010, Indian silk production grew by 2.9 per cent from 13,909 MT to 19,690 MT. Import growth during the period was about 4.76 per cent i.e. 7,530 MT in FY96 to 12,552 MT in FY 10. The overall demand grew by about 3.57 per cent from 21,439 MT in FY96 to 32,152 MT in FY10. However, export earnings grew by 9.82 per cent during the period.

The raw silk production which was around 16,319 MT during the year 2002-03 has increased to 18,475 MT by 2006-07, the end of the Tenth Five Year Plan with an increase of around 13.21 per cent. This happened in-spite of various constraints like drought in traditional sericultural areas of southern peninsula during the years 2002-03 and 2003-04. Additionally, the prices of sericultural commodities had come down during the same period due to large scale dumping of Chinese Silk (yarn and fabrics) into the country.

To support the prices of sericultural commodities, government levied anti-dumping duty on the low grade silk yarn and fabrics imported from China. Currently, the stakeholders of the silk industry are poised to take-up sericulture on a large scale on account of better prices, improved technology and labour saving practices.

Quality of Indian Silk

Improvement of quality of Indian silk has been a matter of serious concern of the Indian sericulture industry. The World Bank aided, Karnataka Sericulture Project was implemented in Karnataka with the main objective of improving quality apart from increasing production. Subsequently, the National Sericulture Project was launched for increasing annual production good quality bivoltine silk yarn by 1000 MT. As a result of Research and Development efforts of the Central Silk Board, number of new bivoltine races has been developed, which has led to improvement of quality of Indian silk. Nevertheless, productivity of bivoltine silk in India has remained considerably low. The average level of Bivoltine Cocoon production in Japan has been around 60 kg per 100 DFL, whereas, the same in India is around

30 kg per 100 DFL. Similarly, the shell ratio is much below the minimum shell ratio (18% achieved in Japan) with percentage of melting is less than 5 per cent. Besides, as per international standard, renditta should be about 7 and quality of silk "2A" grade on average, which are the basic requirements of good quality silk yarn. Quality of Indian silk has to reach comparable standards for surviving the competition from abroad. With the globalization and opening of the Indian economy to outside under the new economic policy and the conditions of GATT, improvement of quality has assumed new dimensions and made quality improvement as a basic requirement.

Research and Development efforts of the Central Silk Board has paid dividends with regards to improvement of quality. An integrated technology package for bivoltine silk production in tropical environment has been evolved by the Board, comprising of the following:

(i) V-1 and S-36 mulberry varieties, together with cost-efficient cultivation practices.
(ii) CSR series of bivoltine hybrids isolated for tropical climate with collaborative effort of CSB and JICA scientists.
(iii) Refined multi-end composite reeling pack, which spreads up to weaving and wet processing.
(iv) Low-cost models of silkworm rearing houses, and
(v) User-friendly silkworm rearing and cocooning equipments.

The Central Silk Board has planned to pursue these with greater vigour for improving the quality of Indian silk in general and increasing production of bivoltine silk in particular.

With a view to improve the quality of Indian silk, production of bivoltine silk has been emphasized time and again under successive plans. Under the National Sericulture Project implemented between 1988 and 1996, special efforts were made for introducing bivoltine silk production in selected districts of twelve non-traditional States namely Assam, Bihar, Kerala, Maharashtra, Madhya Pradesh, Rajasthan, Uttar Pradesh, Odisha, Gujarat, Punjab, Haryana and Himachal Pradesh. However, the achievementsin respect of the bivoltine in the non-traditional states limited to only 10 per cent or the target under the NSP (details given in Table 4.11) were.

Production of bivoltine silk have been emphasized during successive plans. However, the achievements have remained much lower than the target even in the country level. It appears that during the Ninth Plan, production of bivoltine silk increased from 370 MT (in 1999-2000) to 900 MT (in 2001-02). This position improved only marginally during the Eleventh plan with production of only 1685 MT (33.7%) of against target of 5000 MT.

Table 4.11: Promotion of Bivoltine Sericulture in the Non-Traditional States under National Sericulture Project (NSP) (1988-96)

	Target	*Achievement*	*Percentage*
Area under Mulberry (in ha)	8,600	2,565	29.82
Cocoon Produced (in MT)	1,600	160	10
Silk Production (in MT)	160	17	10.62

Source: Central Silk Board.

Research and Development

The Central Silk Board has a network of Research and Development Institutes with a network of Regional Sericulture Research Stations and Research Extension Centers throughout the country to provide necessary research and development support to the Industry. The technology developed by R&D Institutes of CSB have been popularized among the farmers to maximize yield and returns which in turn in bringing out vertical growth of the Industry. Due to R&D intervention the raw silk productivity has increased from 87.84 kg/ha (2007-08) to 90.90 kg/ha (2011-12), Renditta (Nos. of kg of cocoons required to produce 1 kg silk) has reduced from 8.13 (2007-08) to 7.66 (2011-12).

In promotion of sericulture, there is no alternative to achieving quality silk of international standard. The Central Silk Board has started a testing laboratory for Silk and Zari at Kancheepuram to facilitate customers, producers and other stakeholders for spot testing of silk and zari material without any destruction to the products by x-ray analysers. More importantly, 'Silk Mark' Scheme is introduced by Silk Mark Organisation of India (SMOI), a registered society, sponsored by Central Silk Board, Ministry of Textiles, Government of India. The Silk Mark is a quality assurance label for the assurance of pure silk and in addition serves as a brand for generic promotion of Pure Silk.

Significance of Sericulture Industry

Indian Sericulture industry has a history of over thousands of years. Silk has been an inseparable part of Indian culture and tradition as no ritual is considered complete without silk apparel, *sarees/dhotis/shawls*, being used. Silk sarees wear for women, from leading weaving centers, is often regarded as a must for any social or religious occasion. Indian silk were known for their fine quality, lustrous sheen and traditional colours. The luxurious crepes, georgettes, and chiffons of Karnataka, the tie and dye craft of Andhra Pradesh, Gujarat and Odisha, the delicate silks of Kashmir, sheer brilliant fabrics of Bandej and the temple silks of Kancheepuram, along with a host of traditions of sarees and other form of apparels were known worldwide.

In India, sericulture is not only a tradition but also a living culture. It is a farm-based, labour intensive and commercially attractive economic activity falling under the cottage and small-scale sector. It particularly suits rural-based farmers, entrepreneurs and artisans, as it requires low investment but, with potential for relatively higher returns. It provides income and employment to the rural poor especially farmers with small land-holdings and the marginalized and weaker sections of the society. Several socio-economic studies have affirmed that the benefit-cost ratio in sericulture is highest among comparable agricultural crops.

Sericulture is a cottage industry par excellence, with its agricultural base, industrial superstructure and labour intensive nature. It is remarkable for its low investment, and quick and high returns, which make it an ideal industry or enterprise which fits well into the socio-economic factors of India. Comparative cost benefit analysis (details given in Table 4.12) indicate that mulberry sericulture is one of the better option.

Table 4.12: Cost Benefit (CB) Analysis of Mulberry Sericulture with Other Crops

Item	*Mulberry sericulture*	*Sugarcane*	*Turmeric*
Total input costs	48,659	30,575	29,610
Gross returns	96,132	60,200	55,317
Net returns	47,476	29,625	25,707
CB ratio	01:02.0	01:02.0	01:01.0
Crop period	2 months	1 year	4-5 months

Source: unit Rs/acre/annum, Dandin *et al.*, presented at the 20th Conference of the International Sericulture Commission, Bangalore, December, 2005)

Sericulture is recommended as one of the most effective tools for rural development in general and poverty alleviation in particular. Sericulture is a labour-intensive industry in all its phases, namely, mulberry cultivation, silkworm rearing, silk reeling, and other post-cocoon processes such as twisting, dyeing, weaving, printing and finishing. The industry have several other advantages namely– (i) transfer of income from richer to poorer classes, (ii) use of marginal land without affecting foodgrains production, and (iii) high export potential. The industry provides whole or part time employment to more than 5.7 million persons in the rural and urban areas and of these 30 per cent are drawn from the Scheduled Castes, Scheduled Tribes and Other Backward Sections. Till recently, sericulture was considered a subsidiary occupation. Introduction of new technology of sericulture had led to making the industry a highly remunerative cash crop, as reflected in increasing acreage being brought under mulberry cultivation, and the step in the raw silk output being witnessed every year.

Sericulture as a Tool for Empowerment of Women

Women contribute to a little less than 50 per cent of the country's population and most of them are largely rural based, deriving their livelihood through agriculture and other land based activities, either as family members or wage earners. As a cottage industry, sericulture provides ample wok for women in the rural areas particularly in silkworm rearing and reeling, while men, largely work in the field and in weaving. The involvement of women in different activities of sericulture is about 53 per cent and their contribution in the on-farm activities understandably is lower than that in post-cocoon activities. An account of involvement of women in various components of sericulture is given in Table 4.13. Sericulture is an ideal vocation for women because of the following facts:

- Being rural based, sericulture needs less specialized skill and hence, suits women well.
- Involves mostly indoor activities, less physical energy and manual labour.
- Work is evenly spread over during the day, with intermittent gaps, offering leisure time and proximity to living place.
- Minimum investment with long life and short gestation period.
- Continuous job with frequent income and scope for utilizing several by-products for value addition.

Table 4.13: Involvement of Women in Different Sericultural Activities

Sl. No.	*Activity*	*Involvement of women (%)*
1.	Silkworm seed production	20.46
2.	Mulberry cultivation	49.55
3.	Silkworm rearing	49.67
4.	Silk reeling	48.81
5.	Silk spinning	80.00
6.	Silk throwing (twisting)	56.34
7.	Silk weaving	49.02
8.	Dyeing-Printing of silk yarn and fabrics	41.00
9.	Silk bye-products	65.00
10.	Silk and Milk production	75.00
	Average	**53.45**

Source: Central Sericultural Research and Training Institute, CSB, Mysore.

As a sustainable economic activity favouring the rural poor in the unorganized sector, sericulture holds special promise for the women. This is of particular significance as most of the operations such as plucking of leaves and silkworm rearing do not need hard strenuous labour, are done indoor and thus most convenient to the women. Further, women can do this work, while attending to their normal household activities like cooking, looking after the children etc. It has been estimated that various activities on sericulture generate about 13 work years of employment per hectare of mulberry planted. Out of the same, women avail about 60 per cent the total employment opportunity. Sericulture is particularly useful for employment of women, who are illiterate and can not go out of their homes due to social and other constraints. Advancement of women in sericulture has two fold connotations. In the first instance, women participated in group formation and leadership development, group economic activity development, get a say in marketing and input decision-making. As a result, access of women to sericulture income increased. Involvement in sericulture activity enables the women to have a greater say in their own household, community and society in terms of participation in decision-making and control over resources and income. This in turn help in empowerment of the women and removing the gender related bias.

With a view to increase participation of women in silk industry, a number of special steps were initiated namely (i) identification of female-headed households as beneficiaries (ii) inclusion of women in regular extension dialogue, (iii) priority of women and to have a majority of women, among trainees under peripatetic training programmes as well as in Farmers' Training Schools, (iv) support to organizations assisting landless women, Mahila Mandals, female heads of households (v) recruitment of women training staff for providing all assistance training and promotion of women's participation in sericulture, (vi) reservation for women in training programmes, (vii) facility of one time financial assistance for economic group activities of women and (viii) thrust to the developmental efforts for increasing women's participation in sericulture. The year 1994 was observed as the 'Year of women in Sericulture'.

Sericulture and Rural Development

Sericulture being a farm-based enterprise is highly suited for both large and small land holdings, with low capital investment. The very nature of this industry with its rural based on-farm and off-farm activities and enormous employment generation potential has attracted the attention of many planners and policy makers to recognize the industry as one of the most appropriate avenues for socio-economic development of a largely agrarian economy like ours. Generally, silk goods are purchased by the urban rich and middle-class consumers and it is estimated that around 57 per cent of the final value of silk fabrics flows back to the primary producers in rural areas. Sericulture can also play a very vital role in alleviating rural poverty due to its high work participation rate and thereby can check migration from rural to urban areas.

Sericulture and Employment Generation

Mulberry sericulture is labour intensive in nature and is of particular importance for women and the old. An account of employment potential of mulberry sericulture is given in Table 4.14. It is obvious that mulberry sericulture can be an effective tool in solving the problems of unemployment and poverty alleviation in the rural areas.

Strengths, Weakness, Opportunities and Threats (SWOT) of the Indian Silk Industry

R&D achievements like development of indigenous mulberry varieties with highest leaf yields in the world, new bivoltine silkworm hybrids eminently suited to the tropical regions of the country, farmer-friendly technologies, cost-effective new package of practices for cultivation of food plants, rearing and reeling coupled with huge natural and man-made resources and trained manpower clearly indicates the future prospects of sericulture industry to emerge as a promising indicator of economic development for the upliftment of the socially deprived communities and the downtrodden. The Strengths, Weakness,

Table 4.14: Activity-wise Employment Generated in Mulberry Sericulture

(Per Ha.)

Activity	Employment generation Man-days	Man-years
A. Mulberry cultivation and silkworm rearing		
a. Mulberry cultivation	585	
b. Leaf/shoot harvesting	320	
c. Silkworm rearing	350	
Sub-total	1255	5.02
B. Reeling of silk cocoons		
@ 300 mandays per 1000 kgs of reeling cocoons (@ 8.0 renditta; 760 kgs cocon/ha.; 95 kgs raw silk/ha.)	2250	9.12
Total (A + B)	3535	14.14
C. Twisting		
@ 220 g of silk per mandays	432	1.727
D. Weaving		
Handloom @ 0.13 kg per manday	438	1.752
Powerloom @ 0.3125 kg per manday	122	0.486
Sub-total (Weaving)	560	2.238
E. Printing and Dyeing		
@ 40 mandays for 40 kg of raw silk	95	0.38
F. Finishing		
@ 751 mandays for 40 kf of raw silk	1784	7.135
G. Silk Waste Processing		
@ 18.775 mandays per kg of raw silk	26	0.104
Total (C to G)	2896	11.58
Grand Total	6431	26

Source: Central Silk Board.

Opportunities and challenges (SWOT analysis) of Indian silk industry have been given in Table 4.15.

Indian Silk Industry—The Road Ahead

India has the unique distinction of producing all the four varieties of natural silk i.e. tasar, eri, muga and mulberry. It has a sizable

Table 4.15: SWOT Analysis of the Indian Silk Industry

Strengths	*Weakness*	*Opportunities*	*Threats*
Large production base, availability of skills, land and labour.	Gaps in technology transfer and extension support.	Generation of rural employment and reduction of migration to urban areas.	Falling international prices and heavy dumping from China at low prices.
Established infrastructure, availability of silkworm breeds/hybrids.	Inadequate market accessibility, poor linkage among different stake holders.	Liberalization policies of Govt. of India in line with WTO Agreements.	Unpredictability of China's silk policies.
Low investment, short gestation period and higher returns.	Decentralized nature of the industry inhibits financial institute from extending financial support to the sector.	Reduction of production of silk even by traditional silk countries like Japan, USSR etc.	Inability of the silk industry to react and adopt to the changing needs in terms of quality both for the domestic and export markets.
Easily adoptable technologies and strong domestic demand-pull.	Lack of quality based pricing system in the market, frequent price fluctuations and large scale imports from China at low prices.	Garment exports are on a steady increase with huge employment opportunities.	Lack of awareness in the domestic market to respond to the demand-driven milieu.

domestic demand for mulberry silk yarn and fabrics, which is likely to grow steadily with growth of population and general development in the economic condition. At present requirement of mulberry silk is about 30,000 tonnes, of which about 14,600 tonnes is being met from indigenous production. However, bulk of domestic silk is of multivoltine races and relatively low quality. The gap in supply is mainly for better quality silk for use as warp. This constitutes approximately about 42 per cent of the total domestic demand. Warp-grade silk, preferred by power loom weavers can be obtained only from bivoltine races. Imported Chinese bivoltine yarn, of even indeterminate grade, presently fills this supply-gap. The challenges ahead lie in improving quality of domestic silk for exploiting the demand both inside and outside the country.

Production of silk, particularly mulberry silk has seen significant changes over last three decades. In 1987, China topped in production of silk (37,600 tonnes) with India in the second position (9,500 tonnes) followed by Japan (7,900 tonnes) in the third position. Since then, silk production in Japan has gone down significantly to a mere 650 tonnes in 2000. China seized the opportunity and raised its production from 10,200 tonnes in 1970 to about 79,000 tonnes in 1995. China has raised its production from 79,000 tonnes in 1995 to 115,000 tonnes in 2010. The production of silk by India, in the second position is 20,400 tonnes. The silk production in Japan has gone down to a mere 53 tonnes in 2010. The production of silk by South Korea and Brazil is also in the decreasing trend. Production of good quality bivoltine silk has been the major strength of China in achieving this remarkable success. However, production of raw silk in China has been changing, partly due to reprioritization of land-use pattern in China for meeting the requirement of foodgrain and partly due to rapid industrialization leading to diversion of labour from agriculture to the industrial and service sectors. Other silk producing countries like South Korea, erstwhile USSR and Brazil though technologically proficient, are not able to produce silk at internationally competitive prices on account of high labour and other input costs. Thailand and Vietnam are late entrants in to the business and are facing teething problems. These factors indicate that immense opportunity is available for the development of Indian sericulture industry, in the international silk market, in the coming years.

Notes

1. Nanavaty, M.M., *op. cit.*, p. 4.
2. *The Cambridge Economic History of India*, volume I and II, Orient Longman, 1984.
3. *The Gazetteer of Varansai*, Government of Uttar Pradesh, 1962.
4. Report of the Indian Tariff Board, 1933.
5. Royle, J. G., *Essay on the Productive Resources of India*, M.H. Allen & Co. London.

6. Tenth Five Year Plan 2002-07, Planning Commission, Government of India, p. 717.
7. Final report of sub group on Silk-National Fibre Policy, Ministry of Textiles, Government of India.

5

Mulberry Sericulture in New Area

Experience of Odisha

Background

Odisha, located in the eastern coast of India, has a strong traditional textile industry. Handloom woven fabric of Odisha are immensely popular all over the country and abroad because of exquisite workmanship and design. Odisha is also known for production of non-mulberry '*tassar*' silk industry. Large number of tribal families have been rearing tasar silkworm under natural conditions in *sal* and *asan* trees in the natural forests as a traditional occupation. *Tassar* silk worm rearing is common in the Mayurbhanj, Keonjhar, Sundargarh districts adjoining the Deccan Plateau. However, sizeable quantity of mulberry silkyarn required by the handloom industry of Odisha for production of silk garments that is brought from outside mainly from Karnataka. Some quantity of cheaper *Nestari* silk yarn made in Maldah, West Bengal is brought for use in the weft. Besides, better quality *Bivoltine* silk yarn is brought for use as warp.

Keeping in view the requirement of silkyarn for meeting the need of the handloom industry, attempts were made for taking up mulberry sericulture in Odisha. Initially, it was taken up in 32 selected blocks on pilot basis through mulberry demonstration farm (MDF) on experimental basis during the Seventh Plan. The beneficiaries were assisted by various ongoing poverty alleviation programmes. Silkworm eggs and other materials required for the purpose, were supplied by the Central Silk Board (CSB).

Subsequently, CSB set up a grainage for meeting the silkworm egg requirement. Keeping in view similar climate and proximity to West Bengal, rearing of multi-voltine type silkworm was taken up in Odisha at the initial stages.

Odisha was grouped among the non-traditional States so far as mulberry sericulture is concerned. A special project namely Bivoltine Sericulture Development Project (BSDP) was launched in Odisha in 1987-88 with a view to produce high quality bivoltine silk. The project was implemented jointly by CSB and the sericulture wing of the Textile Directorate of Odisha. Subsequently, under the World Bank aided National Sericulture Project (NSP), Koraput district of Odisha was covered for improving quality and productivity of mulberry silk yarn.

Bivoltine Sericulture Development Project (BSDP)

The Bivoltine Sericulture Development Project (BSDP) was sanctioned by the Government of India on July 30, 1986. However, field operation of the project started from April, 1987. The project was implemented with the twin objectives of introducing bivlotine mulberry sericulture on a commercial scale and socio-economic development of the Scheduled Tribe families living in the project area.

The project originally envisaged raising Mulberry plantation in 1000 acres of private lands belonging to 1300 beneficiaries. The plantation was to be raised in both irrigated and rain-fed area, covering 600 units of half (0.5) acre units each and 300 units of one acre each under rain fed condition and 400 units of one acre under irrigated conditions. The project had a target of production of 180 MT of Bivoltine cocoons to be reeled into 17 MT of yarn worth Rs. 76.50 lakhs per annum on completion of the project i.e. from the year 1991-92 onwards. There are 5,000 persons belonging to 1,000 families were expected to be gainfully employed in mulberry sericulture related activities under the project.

The project had a financial outlay of Rs. 427.45 lakhs to be spent on

(i) Development of Infrastructure for supporting sericulture—Rs. 156.55 lakhs to be funded by the State

Government and Central Silk board @ Rs. 78.275 lakh each,

(ii) Provision of required assistance to the beneficiaries—Rs. 157.07 lakhs, to be funded by the State Government Rs. 85.515 lakh and institutional finance (loan) from banks Rs. 71.555 lakh, and

(iii) Setting up Reeling and Twisting Complex in the Co-operative sector—Rs. 113.83 lakhs, out of which Rs. 11.383 lakh (10%) to be funded by the State Government and Rs. 102.447 lakh (90%) by NCDC.

While the cost of infrastructure development was proposed to be shared equally by the State and the Central Silk Board, expenditure for assisting beneficiary was to be borne by the State Government under Special Central Assistance and other ongoing anti-poverty programmes like ERRP/IRDP/ITDA and Loans from Banks.

The project envisaged setting up following infrastructures for supporting sericulture:

1. P-1 Seed Station—one for production of seed cocoon,
2. One Grainage for production of silk worm eggs,
3. Chowki Rearing Centres—10 number for rearing of young silkworm under controlled conditions, and
4. Co-operative Reeling and Twisting Complex. - 2 number.

Odisha State Tasar and Silk Cooperative Society Ltd., an apex cooperative society of Odisha dealing with development of sericulture implemented the Bivoltine Sericulture Development Project from 1st April, 1987.

Achievements

After the project was started with 189 beneficiaries, having plantation in 111 acres in the project area were adopted under the project. Out of them, 164 beneficiaries belong to Scheduled Tribe and 25 to other castes. There were 33 units of 1 acre each and rest were of 0.5 acre each. As additional assistance of Rs. 82,000 was provided to these beneficiaries for development of mulberry plantation, construction of rearing house and rearing equipment.

In the first year of the project 1987-88, mulberry plantation was raised in 108 acres by 135 beneficiaries, out of whom 126 belong to Scheduled Tribes. All these villagers were provided assistance under the ongoing poverty alleviation schemes. There were 108 beneficiaries covered under ERRP, with about 90 per cent government subsidy and 10 per cent own contribution or bank loan. The rest 27 were covered under IRDP schemes with about 75 per cent government subsidy and balance own contribution or bank loan. It was not possible to motivate progressive farmers to take up the practice with their own money. There were 130 rearing houses completed and rearing equipment was supplied to 100 beneficiaries during the year. There were 7705 DFLs brushed and 2076.270 kg of cocoon were produced. The beneficiaries earned Rs. 2,62,360 by sale mulberry cuttings and as wage labour.

In the second year of the project 1988-89, mulberry plantation was raised in 379.5 acres by 496 beneficiaries, out of whom 463 belong to Scheduled Tribes. All the 496 beneficiaries were assisted under ERRP schemes. In addition 11 progressive farmers could be motivated to raise mulberry plantation with their own money. There were 106 rearing houses completed and rearing equipment was supplied to 36 beneficiaries during the year. In turn 11,354 DFLs were brushed and 3077.640 kg of cocoon were produced. The beneficiaries earned Rs. 11,39,832 by sale of cocoon, mulberry cuttings and as wage labour.

In the third year of the project 1989-90, mulberry plantation was raised in 115.5 acres by 141 beneficiaries, out of whom 137 belong to Scheduled Tribes. 121 beneficiaries were assisted under ERRP and 20 under the ITDA scheme. There were 13 progressive farmers, who raised the plantation with their own money. There were 159 rearing houses completed and rearing equipment was supplied to 156 beneficiaries during the year. There were 48,087 DFLs brushed and 15,499.720 kg of cocoon were produced. The beneficiaries earned Rs. 12,10,817 by sale of cocoon, mulberry cuttings and as wage labour.

In the fourth and last year of the project 1990-91, mulberry plantation was raised in 232 acres by 308 beneficiaries, out of whom 287 belong to Scheduled Tribes. All the 308 beneficiaries were assisted under ERRP schemes. There were 20 progressive farmers, who raised the plantation with their own money. There were 273 rearing houses completed and rearing equipment were supplied to 301 beneficiaries during the year. In turn 80,827 DFLs brushed and 17214.150 kg of cocoon were produced. The beneficiaries earned Rs. 14,31,851.30 by sale of cocoon, mulberry cuttings and as wage labour.

On account of several factors, it was decided to extend the project by one more year. In the fifth year of the project (1991-92), mulberry plantation was raised in 128.5 acres by 232 beneficiaries, out of whom 216 belong to Scheduled Tribes. Out of them 198 were assisted under IRDP schemes. There were 34 progressive farmers raised mulberry plantation with their own money. There were 106 rearing houses completed and rearing equipment was supplied to 290 beneficiaries during the year. There were 92,212 DFLs brushed and 25,046.300 kg of cocoon were produced. The beneficiaries earned Rs. 24,19,878.95 by sale of cocoon, mulberry cuttings and as wage labour.

The Study

With a view to assess the impact and various factors critical to success of mulberry sericulture in new area, a pilot study was undertaken in two blocks covered under BSDP. Achievements of the BSDP in the Mohana and R Udaigirai blocks were compared with those for other blocks falling in other NSP as well as other parts of the state. Prevailing conditions were more or less similar in all these areas. However, in the BSDP area, integrated approach was adopted and extension support was provided in intensive manner.

Initially while introducing mulberry sericulture in 32 blocks on pilot basis, rearing of multivoltine and multivoltine-bivoltine hybrid silkworm races were taken up for rearing. Success of bivoltine rearing in the BSDP area indicates that rearing of *bivoltine* race can be introduced straightway in the non-traditional and new area and it is not necessary to go for *multivoltine* race first and then changeover to *Bivoltine* race.

Findings

Sericulture being a labour intensive practice, number of family members is supposed to play an important role in promotion of the practice in general and income from sericulture in particular. In order to analyse the impact of number of family members on income from sericulture, the beneficiaries were grouped into two groups– (i) families having less than five persons, and (ii) those having five or more persons. It is found that 67 families, having less than 5 members had an average income of Rs. 1953 and on the other hand, 93 families having five or more members had an average income of Rs. 2163 during the year 1991-92. The fact that earning from sericulture per year for families with five or more members has been higher than families with lesser members.

Education has been an important contributing factor in income soufces. To see the role of level of education of the beneficiary on income from sericulture, the respondents were divided into four groups namely– (i) illiterate, (ii) literate, (iii) educated up to primary level, and (iv) educated above primary level. It has indicated that the illiterate group had the highest income of Rs. 2164 per acre, followed by the literate beneficiaries. Beneficiaries, who had up to primary level of education, had an average income of Rs. 1232 where as those who had education above primary level had a higher income of Rs. 1580. These facts indicate that education level of the beneficiary do not contribute in increasing income from sericulture.

Similarly, on analyzing the impact of age of the head of family on income from sericulture per acre during 1991-92, beneficiaries were divided into three groups, namely– (i) those, who were less than 30 years old, (ii) those, in the 30 to 40 years age-group, and (iii) those, more than 40 years old. It is found that the average income of these three groups of beneficiaries was to be Rs. 2164, Rs. 1771 and Rs. 2163 respectively. Hence, it is found that the relatively young and old beneficiaries had income higher than the middle group.

With regards to land held by the beneficiary on income from sericulture, it is found that the farmers having 1 to 2 hectares of land had the highest income of Rs. 2216. Those having 2 to 5 hectares had an average income of Rs. 2104, whereas those having less than one hectare of land had an income of Rs. 1973 and those having more than five hectares had the least income of Rs. 1752.

The training of the beneficiary on income from sericulture has given a positive thrust and it is found that there is a positive correlation between training and income from sericulture. During 1991-92, those, who had not been trained at all, had an average income of Rs. 1488, which is the lowest income. Those, who had been trained locally, had an income of Rs. 2021, which was the average income. In case of those, who were trained and also had the benefit of visiting sericulturally advanced States, had the highest income of Rs. 2696, which is about 30 per cent higher than those, who only training and about 80 per cent higher than those, who had no training at all.

A separate house for undertaking rearing of silkworm is an essential requirement for success of mulberry sericulture. In case there is no separate house for rearing, the rearer had to take up rearing in his own dwelling house. In such a case, it is not possible to take up disinfections, which makes the silk worm more prone to diseases during rearing. From the study, it appears that rearers, who did not have a separate house for rearing, had an average income of Rs. 1698. Average income of rearers having a separate rearing house was Rs. 2134, which was about 25 per cent higher than who did not have a separate rearing house.

Like the rearing house, availability of required rearing equipment for undertaking rearing of silkworm is another essential requirement for success of mulberry sericulture. In case there is no separate equipment for rearing, the rearers had to take up rearing either on the floor of the house or otherwise. In some cases the rearers have only part of the required equipment. This leads to crowding of worms and the quality and productivity are affected. In such a case, it is not possible to take up disinfections properly and the worm becomes more prone to disease. With a view to ascertain impact of availability of rearing equipment on income from sericulture during 1991-92, it is found that those having only part of rearing equipment had an average income of Rs. 1686. This was about 30 per cent lower than Rs. 2776, average income of rearers, having required rearing equipment in full.

Most of the beneficiaries covered under the project belong to Scheduled Tribe and were having income below the poverty line. They did not have adequate means for taking up mulberry sericulture of their own. As such the beneficiaries were assisted under ongoing schemes for alleviation of poverty. Under the programme, Economic Rehabilitation of the Rural Poor

(ERRP), implemented by the State Government of Odisha, ST families were provided assistance up to 90 per cent for various economic activities. Under the Integrated Rural Development Programme (IRDP), eligible families were provided assistance up to 75 per cent. With a view to encourage relatively well off persons for taking up sericulture, another scheme was introduced under which assistance up to Rs. 4000 was provided for taking up sericulture in one acre. This works out to about 23 per cent of the total fund required. In the study under the project it found that those covered under ERRP had got an average income of Rs. 2244, which is the highest income. Those covered under IRDP had an average income of Rs. 1289, which was about 57 per cent of the income of the first group. Families covered under the progressive beneficiary scheme had an average income of Rs. 1390, which was 61 per cent of the income of the first group. However, income of the so-called progressive beneficiaries was higher than the beneficiary assisted under IRDP.

Income and Employment Potential of Mulberry Sericulture

One of the most important features of mulberry sericulture relates to relatively high economic return with low capital investment. Return from sericulture mainly comes from sale of cocoon, which is reeled to silk yarn. Sale of mulberry cutting, obtained from pruning the old plantation, supplements the income to some extent. Under irrigated condition, rearing is done continuously almost throughout the year, which gives steady income to the rearers. Beauty of mulberry sericulture lies in the fact that silk is a costly fabric and is purchased by the rich. Thus, mulberry sericulture acts as a medium for transferring wealth from the rich (consumer of silk fabrics) to the poor (the producers engaged in plantation of mulberry, rearing of silk worm, reeling and twisting of silk yarn and weaving silk fabrics). Share of various segments of the silk industry, given in Table 5.1 indicates that silkworm rearing has the highest (56.8%) share out of the total value of silk fabrics. This aspect of sericulture makes it an ideal tool in ensuring growth with distributive justice.

As per the State specific poverty line for Odisha, Rs. 121.42 was the financial limit for deciding poverty line in rural areas

Table 5.1: Share of Various Producers in the Sericulture Industry

Sl.	*Activity*	*Return (in Rs per acre)*	*Share of the total return (percent)*
1.	Silk Rearer	8724	56.8
2.	Reelers	1044	6.8
3.	Twister	1397	9.1
4.	Silk Weaver	1643	10.7
5.	Trader	2550	16.6

Note: Balasubramanium, V., *Indian Silk*, Oct 1986 (based on the assumption 12.8 kg of silk per acre, 15 meters of silk fabrics per kg silk, Rs. 80 per meter of silk fabrics.)

for the year 1987-88, which was subsequently revised to Rs. 194.03 for the year 1993-94. As no separate poverty line figures for 1991-92 was available at the time of study, same was deduced as Rs. 163.14, based on the wholesale price index figures as applicable to the project area. This corresponds to Rs. 1957.68 (say 1958) per capita per year. While persons having income less than the poverty line are taken as those living below poverty line, those having income less than 50 per cent of the said line are taken as ultra-poor. Analysis of the data of income relating to sericulture for the surveyed beneficiaries is given in Table 5.2:

Table 5.2: Income from Sericulture with Reference to Poverty Line

Income from sericulture (in Rs per member per year)	*Number of beneficiaries*	*Percentage of beneficiaries*
1-979 ultra poor (< 50% of poverty line)	137	85.6
979-1958 poor (>50% of poverty line but < poverty line)	19	11.8
>1958 (above poverty line)	4	2.6

Return to individual rearers from mulberry sericulture in the Project area indicates that it varies considerably due to several factors. Shri Jakshya Gomango of Gopalpur village got Rs. 8888 from mulberry sericulture in one acre under rainfed condition, which was the highest of all beneficiaries. Data for the surveyed beneficiaries given in Table 5.2 indicates that in case only income from sericulture is considered, only 2.6 per cent of the beneficiaries had income above poverty line and 11.8 per cent beneficiaries had income higher than half of the poverty line in the year 1991-92. In other words, return from mulberry sericulture alone under the project was not adequate enough to enable a family to cross poverty line.

With regards to the poverty alleviation, mulberry sericulture in the project area has contributed a lot towards income stability of the beneficiaries who were living below the poverty line and to the nomadic tribes. Prior to starting of mulberry sericulture, the local villagers were earning their livelihood from traditional agriculture and shifting cultivation practiced in the hill slope. It has shown that after mulberry sericulture was introduced, income from sericulture accounted for about 67.5 per cent of the total income of the selected families. Income from paddy cultivation and hill slope cultivation (shifting cultivation), accounted for a much smaller component of the total income i.e. 11.64 and 11.85 per cent respectively. The new income pattern also indicated that taking up of mulberry sericulture has succeeded in converting the nomadic tribes into settled agriculturists, which is a major achievement in changing life style of the tribal introduced to sericulture.

Similarly, the comparative study indicate that prior to the implementation of the project in 1987-88, about 94 per cent of the beneficiaries were living in poverty and 59 per cent of them were ultra poor those who were living in acute poverty conditions with income less than half of the poverty line. Subsequent to implementation of the project in 1991-92, about half (49%) of the families had income above the poverty line. Only 9 per cent of them continued to be ultra poor those who have income below half the poverty line and rest 42 per cent had income less than poverty line but more than half the poverty line.

Comparing sericulture with other occupational cultivation in BSDP area it has seen that mulberry sericulture generates higher income than other land based occupations. The net benefit to cost ratio in case of mulberry sericulture was 1.63, which was much higher than other land based occupations like paddy, coarse grain (ragi), horticulture (banana, mixed fruits) and vegetables. The details are given in Table 5.3:

Table 5.3: Income and Expenditure from Sericulture and Other Economic Activities

(*amount in Rs. per Acre during 1991-92*)

Activity	*Gross income*	*Cost of cultivation*	*Net income*	*Net Benefit to cost ratio*
Paddy	1135	628	507	0.80
Ragi	960	595	365	0.61
Banana	2215	1225	990	0.80
Mixed fruit	2692	1350	1342	0.99
Vegetable	1775	965	810	0.83
Sericulture	6800	2585	4215	**1.63**

As per a socio-economic survey conducted in the project area in 1981, household income of the Scheduled Tribes from all sources was Rs. 1528.15 per year. As a result of implementation of various developmental schemes relating to horticulture, animal husbandry, etc., average income of the beneficiaries household increased by Rs. 104.60 by 1987-88, which implied an increase by 6.84 per cent only. As against this, beneficiaries survey have been found to have an additional income of Rs. 2095.38 per household sericulture from sale of mulberry cocoon, which indicate a much higher performance.

Employment Potential of Mulberry Sericulture

Apart from income, generation of higher employment at a relatively low capital investment is another important feature of mulberry sericulture. It is of special significance for development of the rural areas and the women, old and young in particular. Sericulture generates direct employment in mulberry plantation, silkworm rearing and reeling of cocoon. Besides, it generates indirect employment in various service related activities such as construction of rearing house, preparation of rearing equipment, transportation of cocoon, stifling and storing of cocoon, twisting, dyeing, weaving, trading, etc. due to these features, mulberry sericulture provides solutions to the problems of under employment and hidden unemployment and serve as an ideal tool in poverty alleviation.

To see the employment potential, the studies conducted in

the traditional states and seen that various activities relating to mulberry sericulture in one acre under rainfed condition generates about 116 man-days of employment in a year. Out of the various components, rearing of silkworm is the most labour intensive activity and generates about 62 per cent (72 man days) of the total employment. Details are given in the Table 5.4.

Table 5.4: Employment Potential of Mulberry Sericulture Under Rainfed Conditions (1991-92)

	Operation	*Man days/acre/year*
A. Mulberry plantation (initial)		
1	Land preparation	73
2	Plantation	6
3	Application of manure/chemical fertilizer	3
4	Intercultural operation	15
	Total	**97**
	Per year (with 15 years re plantation period)	6.46
B. Maintenance of mulberry garden		
1	Application of manure/chemical fertilizer	4
2	Intercultural operation	16
3	Weeding	9
4	Pruning	5
	Total	**34**
C. Silkworm rearing		
1	Young age rearing	25
2	Late age rearing	32
3	Spinning and harvesting of cocoon	15
4	Marketing	4
	Total	**76**
	Grand Total (A+B+C) 116	

The project had 963 acre land under mulberry plantation in rainfed condition during 1991-92. An account of employment generated in various components during 1991-92 was collected and has been compiled in Table 5.5.

Table 5.5: Employment Generated in the BSD Project Area During 1991-92

Sl.	*Operation*	*Man days*
1.	Maintenance of plantation	9635
2.	Silkworm rearing at chowki stage	4933
3.	Silkworm rearing at adult stage	60480
4.	Cocoon collection, stifling, storing etc	1252
	Total	**76300** (79 per acre)

From the above data it appeared that 79 man-days of employment were generated per acre during 1991-92 in the project area. This works out to 68 per cent of the employment generated in the traditional states.

So far as the use of family's own labour and hired labour is concerned, it is seen that the beneficiaries mostly used their family labour. Studies conducted in the traditional states indicate that for raising mulberry plantation, 15 mandays of labour is hired as against 19 mandays contributed by the family and ratio of hired labour to family labour comes to 0.79. However, rearing of silkworm is done mostly by the family labour with contribution of 66 mandays and required only 10 mandays of hired labour. The ratio of hired labour to family labour in case of silkworm rearing comes to 0.15. These indicate that cost of input and family labour play an important role in mulberry sericulture. Silk worm rearing was being done indoor and mostly by women, whereas plantation related activities was done outdoor and required strenuous labour. There is increased dependence on family labour and hired labour is least preferred and the quality and nature of employment generated by mulberry sericulture is more suitable to the rural folks and particularly the women.

Importance of Linkages in Promotion of Mulberry Sericulture in New and Non-Traditional Areas

Increase in production of cocoon by the rearers and overall success of the project was attributed to the hard and sincere work put in by the project staff particularly those doing the field level extension work to a great extent. The staff provided on the job guidance to the tribal rearers in the rearing technology like raising mulberry plantation, harvesting leaves, feeding the worm, cleaning bed etc. Notwithstanding the in hospitable climate and physical hardship, they provided extension support, which helped in grounding of the project. The project area was very prone to malaria and most of the staff had several round of attack of malaria. Facilities for education, health, etc. were very poor. Excellent work put in by the field was amply proved by the increase in production of cocoon by the rearers and overall success of the project.

Project area, which was new and non-traditional to mulberry sericulture, became self sufficient in availability of plantation material and mulberry cutting by the third year. In subsequent years, the old plantation provided adequate plantation materials for raising new plantation even in the nearby areas. Sale of mulberry cutting gave a subsidiary income to the old rearers.

Mulberry saplings were raised for use in new plantation in the initial years. However, this was not followed up in the later years. In view of the advantages of raising plantation with mulberry sapling, preparation of sapling needs to be given attention. Enterprising rearers, particularly women may be encouraged and assisted for setting up nursery for raising mulberry sapling in a decentralized manner and advantages of raising plantation with sapling should be publicized. This would facilitate raising plantation with sapling, getting better plantation, higher yield of better quality mulberry leaves and higher return to the rearer.

Provision of required infrastructure like seed farms, grainage for quality seed production appeared to be a creditable achievement of the project. However, capacity requirement of silkworm seed was overestimated. This was partly due to the fact that mulberry plantation could not be raised only in irrigated land as proposed initially. Further, there was time over run as time projected for setting up the infrastructure appeared to be under estimated.

On completion of the project, two P-1 seed stations were set up at Ramgiri and Akili for producing about 20 lakh seed cocoons. The balance 20 lakhs seed cocoons was planned to be purchased from the selected seed-rearers. One commercial (F-1) Grainage, was constructed at Chandiput with a capacity of 10 lakhs Disease Free Layings per annum. The grainage was provided with a cold storage for consignment of layings. Taking about 300 DFL per acres under rainfed condition, about 3 lakh layings were required annually in the project area and balance 7 lakh DFLs were earmarked for meeting need of the other parts of the State. Thus, the project ensured assured supply of seed material, which is an important linkage for success of sericulture programme.

Grainage set-up under the project had adequate capacity to cater to the demand of the neighbouring states, after meeting the requirement of Odisha State. However, capacity utilization was low and expenditure incurred for running the grainage remained too high as compared to the earnings. It appeared that it would be difficult to make the grainage commercially viable of its own and part of the recurring expenditure may have to be supported by the Government.

Supply of chowki worm in place of layings to the rearers was useful in taking up rearing in a non-traditional area, particularly in checking incidence of diseases and giving a higher yield. However, chowki reared worm was being given free of charge to the rearers under the project and only cost of laying was being recovered, while purchasing the cocoon by adjusting it against the cocoon value. Cost of preparation of chowki worm was being subsidized on 100 per cent basis under the project. In the initial stage, such an approach was taken for promoting the activity in a non-traditional area and particularly with rearers, belonging to Scheduled Tribe. However, continuing such support on a long-term basis may not be feasible. On the other hand, such subsidy appeared to have adversely affected the initiative taken by some rearers in undertaking chowki rearing in the private sector. This aspect was needed to be given attention and chowki rearing had to be undertaken in the private cooperative sector on a marginal profit basis for sustaining mulberry sericulture activity on a long-term basis.

Supply of other inputs than chowki worm like formalin for disinfections, net for use in the cleaning of rearing bed, Resham Kit Ousadh, etc., on credit through the primary cooperative society to the rearers and recovering the cost from price of cocoon which appeared to be a useful service to the rearers and was self-sustainable. Such approach may be adopted in other areas.

Experience of the project in collection of harvested green cocoon from the beneficiary growers, stifling the cocoon in 'Usmakotis' (operated by hot air made by burning mulberry twigs), storing the stifled cocoon temporarily proved to be useful. This was of considerable significance in reducing loss incurred in processing cocoon, while introducing mulberry sericulture in non-traditional area.

Orgainisation of Primary Mulberry Rearers Cooperative Society for undertaking these operations and providing marketing arrangements was an interesting experience, which is worth emulating in non-traditional areas.

Coordinating production and sale of Bivlotine seed cocoon in an organized manner through the State Silk Society appeared to be a good practice as demand for seed cocoon was sporadic. It was particularly useful in overcoming the expectation of the rearers for getting value for commercial reeling cocoon at a rate equal to that of seed cocoon, notwithstanding the fact that it is often not possible to sale the entire quantity as seed cocoon due to inadequate demand. Similar practice can be replicated while introducing sericulture in new area.

Purchase and reeling operations had considerable risk and uncertainty in areas, where sericulture was introduced for the first time. Available margin in this operation was low. Bivoltine incentive scheme was linked up with actual production and was innovative in nature and proved to be a timely and positive intervention for supporting introduction of mulberry sericulture in new areas in general and the production of bivoltine cocoon and yarn in particular.

The State Silk Society played a critical role in introduction of mulberry sericulture in the project area, both in implementing the project at the initial years and in providing forward and backward linkages for sustaining the operation after the project period. It is necessary that management of the society should be given due attention.

Necessary guidance and support had to be extended by the Government for enabling the apex silk society to carry on its activities in an economically viable manner. Financial support for activities entrusted to the society, which were not economically viable and promotional in nature, encouraging democratic spirit and mass participation of the members, taking steps for reduction of establishment related expenditure are some of the issues, which need consideration in this regard.

Achievements at Macro (Project) Level

Prevailing insecurity about supply of foodgrains, which is typical to the tribal dominated interior area, appears to be one of the major factors behind the villagers growing paddy, millets and other coarse grains for getting the food grains in available land and raising mulberry plantation only in the marginal land. Advantages of mulberry sericulture in general and return to successful rearers in particular are required to be publicized through mass media for overcoming this situation.

Keeping in view, the limitation of irrigation facility, improved and appropriate technology like drip irrigation, pot irrigation by using family labour along with water conservation techniques like mulching, should be introduced in the area for raising mulberry under irrigated conditions. Apart from increasing return to the farmers substantially, this will help in stabilizing the practice and introducing appropriate technology.

Production of quality seed cocoon and grainage operation indicates the success of the project in empowering the beneficiaries belonging to weaker section in taking up a technologically intensive activity on a viable basis.

All villagers interested for taking up mulberry sericulture in a particular village should be assisted either under the ongoing schemes or with interest free loan, for construction of rearing house and rearing equipment. The loan may be recovered in small installments from the sale value of cocoon.

Experiences of the project indicated that the rearers continued the activity because of the assured marketing support. While there may be debate over the rate paid for the cocoon produced, it is beyond doubt that the assured marketing provided by the State Society, played an important role in stabilizing the operations particularly in the initial stages.

In a new area, quantity and quality of cocoon produced would remain uncertain for quite some time both on account of the practice being new and as mulberry is raised under rain fed conditions. As such, arrangements for stifling and storing cocoon was important for avoiding loss of cocoon due to emergence of the moth and also for ensuring regular supply to the reeling unit for making it viable. Experience of the project indicates that stifling green cocoon in the indigenous technology (Chinese) based *Usmakoti* and storing the stifled cocoon locally for some time till it could be transported or reeled played a very important role in reducing loss and making the operation viable.

Reeling of cocoon is one of the weakest and most important links in sustaining sericulture activity in a new area, as not only margin in reeling is low but also it is very risky also due to the perishable nature of the commodity (green cocoon). Experiences of the project indicate that in a new area the villagers, particularly women can pick up the skill in reeling very fast. This can be done in three stages, reeling the low grade cocoon with advanced charkha, followed by reeling with country basins and finally reeling better grade cocoon with improved multi end basins. Management of reeling unit by Rearers cooperative done in the project area offers a good example. Limitation of reeling on account of perishable nature of cocoon and limited margin should be discussed with the rearer members in details for demystifying the cocoon purchase rate and viability of the entire operation. The project indicates that combining reeling with twisting and weaving help in making the overall operation viable due to cross subsidization.

The details of activities of the Cooperative Society, particularly the profit and loss of the society, manner in which purchase rate of cocoon was being decided, etc. were not discussed in the society. An average member got an impression that prices are fixed by the Government and can be increased further. The rearers did not appreciate commercial realities, particularly difficulties faced in introducing sericulture in a new area. Further, during visit to sericulturally advanced states like Karnataka, the rearers got a feeling that market rate of cocoon there was much higher than the rate given in the project area. This led to an impression of being exploited among the rearers and affected their active involvement in the affairs of the society. Problems faced in marketing of cocoon need to be demystified for ensuring greater participation of ordinary rearers in activities of the cooperative society.

The training, particularly organized locally in the demonstration form, was quite useful in passing the skill and increasing productivity. Apart from giving induction training to all rearers, intensive training should be provided to the lead rearers for acquiring new and appropriate technology periodically. So far as visit to sericulturally advanced states like Karnataka is concerned, it was found that while such visits was useful in creating awareness about the programme, it had the adverse effect of rearers expecting a higher rate for the cocoon in the project area, at par with the prevailing rate in Karnataka. With a view to avoid the adverse impact, senior officers should interact with the rearers, before and after their visit and make them understand about the constraints of introducing mulberry sericulture in a new area, where required infrastructure and market are yet to develop.

Non-Government Organisations (NGOs) dealing with welfare of women and Scheduled Tribes, was suggested to be involved particularly for motivation and training. Interested NGO may also be assisted for taking up chowki worm rearing and supply and marketing of cocoon on a "no profit no loss" basis.

Provision of Farm Yard Manure/compost enhances productivity of mulberry plantation. This also helps in disposal of wastes and retention of moisture in the soil. Preparation and use of organic manure may be stressed and pursued with involvement of NGOs.

Till mulberry sericulture is raised in irrigated land and rearing is taken up on continuous basis, problems of cattle grazing will continue. Green fencing provides an economic way of checking cattle grazing. This may be pursued and plantation material may be made available and the rearers may be encouraged to contribute at their own labour.

Intercropping of food crops along with mulberry in the initial stages help in generating additional income and suggested to be pursued in new areas.

Keeping in view the need of quality seed as one of the most important pre-requisites for success of mulberry sericulture, grainage operation should be supported instead of dealing with it only from a commercial and profit making angle.

So far as the reeling complex and Chowki Rearing Centres are concerned, such activities shall have to be passed on to private entrepreneurs in due course. Experience of the project indicates that due to location in interior area and limited production and profitability, very rarely persons from outside are interested for coming to such area. As such interested local entrepreneurs may be assisted and actively promoted for taking over these activities on commercial lines. This is of utmost importance for sustaining sericulture in a non-traditional area.

Availability of land for the setting up the infrastructure required in new area was a major constraint as the area fall under forest and scheduled area. As a lesson for future, necessary permission for land may be provided while approving the project itself.

Even though the Extension staff of the Project had succeeded in convincing the local villagers about sericulture, most of the villagers who were interested to take up sericulture, were rejected as for assistance/bank loan on account of default with the banks or owing to the coverage in other programmes. Despite best efforts to concentrate in a few villages, the number of villages increased. As a result rendering technical support and monitoring became extremely difficult. As a lesson from the case study, mulberry sericulture should be introduced on village based geographical clusters and all villagers living in the area and interested for taking up mulberry sericulture, should be assisted for taking up this culture in a cost effective manner.

Conclusion

The study of achievements of the special project (BSDP) in Odisha provides several interesting insight to introduction of mulberry sericulture in a new and non-traditional area. This project not only envisaged introduction of mulberry sericulture on commercial scale, but also sought to take up rearing of *bivoltine* breed of silkworm for production of high quality silk yarn. Rehabilitation of Scheduled Tribe families is another important feature of the project, which is socially significant. Some of the major findings of the study are given below, which would be of considerable help in expansion of mulberry sericulture in new and non-traditional areas in coming years.

- While introducing mulberry sericulture in new area, it should be taken up in geographical clusters which will facilitate provision of forward and backward linkages and support from institutions like grainage for production of quality yarn, egg and arrangement for reeling and marketing of cocoon.

- In the traditional silk growing areas, the return from sericulture is confirmed which motivates the farmers to take up sericulture under irrigated condition. Under irrigated conditions, several rounds of rearing are taken up almost continuously throughout the year under controlled conditions. This enabled the rearers to get relatively high and continuous yield of cocoon and give economic return. However, in new areas as return from sericulture is yet to get established, very often farmers show reluctance in taking up sericulture in irrigated land which gives a steady income from the traditional crops like vegetables. As such, in new areas mulberry sericulture is essentially limited to rain fed conditions. Under such condition, cattle grazing are a major problem as the existing cropping pattern is limited to a single crop linked up with monsoon. Taking up of mulberry sericulture in rain fed condition in isolated plots becomes difficult to sustain on account of protection from cattle grazing. As such, in a new area all the farmers living in a selected village and having land in contiguous plots should be encouraged to take up mulberry sericulture. For the purpose, depending upon the economic status and eligibility under poverty alleviation schemes, the farmers may be assisted partly by the government and the remaining in form of loan under easy terms.
- Mulberry sericulture is labour intensive in nature. Bulk of the labour is required for rearing silkworm which is done indoor and can be best done by women and elderly persons. Women can do it along with attending to other household works like cooking, taking care of the children. In the non-traditional area, mulberry sericulture is easily linked up with poverty alleviation programme with women members as the focus. Experience of Odisha has confirmed provision of employment and income to the families below poverty line. It has been found that income from sericulture has remained steady in respect

of families having more number of persons who have been able to take up rearing of silkworm and get a higher income.

- The level of attention of the households and training imparted has a strong correlation with return from sericulture.
- Provision of a separate house for rearing silkworm and equipment used for rearing silkworm have a determining role in return from sericulture. Provision of adequate assistance for construction of rearing house and full rearing equipment sets need to be provided for success of sericulture.
- Preparation of sapling for raising plantation has a positive impact in stabilizing plantation earlier and return from sericulture. Preparation of sapling can be best done by decentralization with active performance of the farmers rather than developmental units.
- Rearing of young age worm has a positive role in stabilizing sericulture in new area and reducing instances of disease. The selected lead farmers who are more enterprising may be actively pursued for taking up chowki rearing as a commercial activity. Rearing and supply of chowki worm instead of silk worm eggs can enhance productivity and return from sericulture.
- Organisation of the rearing lies into cooperative which has a strong positive impact in stabilizing sericulture in new area. The rearing cooperatives can play a very important role in supply of egg, chowki worm, other inputs required for rearing as well as marketing of cocoon.
- Arrangement for programme for stifling and reeling of cocoon is critical for success of mulberry sericulture. Indigenous technology based ushnakoti can be effective in stifling green cocoon in new area which prevents margins and losses. Arrangement for stifling and storing cocoon is useful in reducing loss and providing raw material to the reeling unit on continuous basis.
- A price incentive for production of bivoltine cocoon was started, but it has a marginal role. It would emerge better to stabilize the arrangement for purchasing cocoon. On

the one hand, increased productivity of cocoon and on the other hand, for stabilizing sericulture. Selling part of the bivoltine cocoon as seed cocoon is helpful in providing remunerative price to the primary growers in the initial stage. At the state level, suitable mechanism is required to be evolved in providing required attention to sericulture particularly provision of linkage for supply of inputs and marketing which would become important to stabilize sericulture.

- Appropriate technologies like drip irrigation are useful in increasing productivity.

6

Mulberry Sericulture in New Area:

Experience of Tripura

Background

Tripura is a small hilly and landlocked state of the North Eastern region of India. Economy of the state is primarily based on agriculture and allied activities, which provide livelihood to about 70 per cent of the population. The state has a geographical area of 10,497 sq. km., which is predominantly hilly. It is surrounded by deltaic basin of Bangladesh on three sides except for a small part in adjoining Karimganj district of Assam and Mamit district of Mizoram.

Located between 22.57° N and 24.82° N latitudes and between 91.10° E and 92.22° E longitudes, Tripura has the Tropic of Cancer passing through it. Terrain of the state is by and large hilly, with small hills and hillocks criss-crossing the valleys. The State has a typical monsoonal climate ranging from sub- tropical to temperate conditions in hilly areas. Climate of Tripura exhibits a strong seasonal rhythm. There are four seasons in a year; winter (December–February), pre-monsoon (March–April), monsoon (May–September) and post monsoon (October–November). The monsoon normally lasts for about 5 months from May to September and is the longest season of the state. The average annual rainfall of the state is around 2100 mm. The maximum and minimum temperatures during summer are 38° C and 25° C respectively while that of maximum and minimum during winter is 25°C and 7°C respectively. The relative humidity remains as high as 90 per cent during monsoon. The soil has medium fertility. It varies in reaction from very strong to strong acidic with medium organic matter, low phosphorous and potash content. The pH varies from 4.8 to 5.8. The soil texture varies from sandy clay to clay loam.

With about 60 per cent of the land declared as forests, the state has one of the highest green coverage. In the North Eastern (NE) region, Tripura is the second smallest in size but has second largest population. As a result there is high pressure on land. The extent of industrialization has been low due to geographical isolation, inadequate infrastructure, transport bottleneck etc. The land holdings are small and incidence of poverty and unemployment are quite high. Good soil, rainfall and unemployment both direct and hidden, make the conditions conducive for development of land based economic activities including sericulture.

Introduction of Mulberry Sericulture in Tripura

There are references to sericulture being practiced before independence during the princely regime, as evident from a place named "*Resham Bagan*" (Silk garden) located near the Agartala city. After Tripura got statehood in 1972, mulberry sericulture was introduced on demonstration basis with the setting up of Mulberry Demonstration Farms (MDF) as the nucleus. Initially sericulture was limited to rearing of *Nistari* variety of mulberry silk worm race, which was typical to the Malda and Berhampur regions of West Bengal. Sericulture practice at individual rearer's level was started during the Fifth five year plan with support from the North Eastern Council. This was extended in the subsequent period by assisting interested beneficiaries.

Tripura Women Sericulture Project (TWSP)

Launching of the Tripura Women Sericulture Project (TWSP) in 1995 marked the beginning of commercial sericulture in Tripura. With an outlay of Rs. 4.37 crores, TWSP was sanctioned by the Ministry of Women and Child Development for empowerment of women through mulberry sericulture. The project has following major components:

(i) Generation of awareness about advantages of mulberry sericulture.

(ii) Organising women beneficiaries into groups at the village and block level.
(iii) Capacity building by training the field staff and beneficiaries for taking up sericulture on a commercially viable manner.
(iv) Assisting the beneficiaries for taking up good quality mulberry plantation, rearing of silk worm.
(v) Setting up of infrastructure for purchase, storage, reeling of cocoons, twisting of silk yarns, conversion of silk yarns into fabrics and marketing the fabrics.
(vi) After completion of the project period, existing rearers, who had taken up mulberry sericulture under the project, as well as new rearers were assisted under schemes of the Central Silk Board and State Government for pursuing mulberry sericulture. With a view to assess the impact and ascertain factors responsible for promotion of mulberry sericulture in a new and non-traditional state like Tripura, various aspects have been examined.

Mulberry Plantation

In Tripura, mulberry plantation was taken up with the *Mysore local* variety in 1976. Under the TWSP, high yielding varieties of mulberry like K_2 *and* S_1 were introduced. Subsequently, number of other varieties of Mulberry including $S_{1635,}$ V_1 were introduced. An account of mulberry varieties cultivated in the State is given in Table 6.1.

Table 6.1: Characteristics of Mulberry variety planted in Tripura

Variety	*Period*	*Experience*
Mysore local	1976-1995	Hardy, low productivity
K2	1995-2004	Hardy, low productivity
S1	1995-2004	Hardy, low productivity
S1635	2004-2010	Hardy, high productivity
V1	2010	Experimental stage

Preparation of Mulberry Sapling

Initially, Mulberry plantation was being raised with mulberry cutting. Undertaking mulberry plantation with sapling was started under the TWSP. Mulberry sapling was prepared in the departmental farms with government fund by the government employees. Subsequently raising of mulberry sapling was taken up by the Self Help Group(SHG)s as well as individual farmers under Decentralised People's Nursery(DPN). An account of mulberry saplings prepared during 2007-08 to 2011-12 is given in Table 6.2.

Table 6.2: Raising of saplings

(Number of Saplings)

Centres/Year	*2007-08*	*2008-09*	*2009-10*	*2010-11*	*2011-12*
Govt. Farm	2,10,000	2,85,000	3,15,000	2,70,000	1,82,500
Self Help Groups	50,000	1,25,000	3,00,000	2,35,000	12,77,500
Individual Farmer	1,65,000	1,00,000	2,00,000	3,45,000	21,90,000
Total	4,25,000	5,10,000	8,15,000	8,50,000	36,50,000

Number of mulberry saplings raised increased from 4,25,000 in 2007-08 to 8,50,000 in 2010-11 and to 36,50,000 in 2011-12. Saplings raised by the individual farmers under the decentralised people's nursery increased at a much higher rate than in the departmental farms. Analysis of the trend indicates the steep increase during 2011-12 mainly by the individual farmers. Dovetailing MGNREGA fund for preparation of sapling appears to be a major factor behind this trend. Raising plantation with sapling appeared to have helped in stabilising the mulberry plantation early and giving a higher yield of leaf and return to the rearers.

Experiences of Tripura indicate that while introducing sericulture in a new area, it would be prudent to raise mulberry plantation with sapling, prepared by individual farmers and SHG under decentralised people's nursery (DPN) rather than in the nursery managed by the department with government fund. Using MGNREGA fund for raising sapling appeared to be a good example of using MGNREGA fund for productive purpose.

Silkworm Rearing

Initially *Nistari* breed of silkworm, common in the Malda area of West Bengal, was being reared in Tripura till 1995. Notwithstanding the inferior quality, poor yield and return, rearing of *Nistari* breed continued as it was considered relatively hardy and less prone to silkworm disease. Lack of basic facilities like a separate rearing house, facility for disinfection, inadequate training of rearers were forcing the rearers to remain satisfied with the low productivity of about 20 kg of cocoon per 100 Disease Free Laying (Dfl)s reared. Launching of TWSP in 1995 marked the beginning of a new era with the introduction of high yielding crossbreed races being reared in Karnataka. Over years, rearing of the high yielding breeds has become popular and *Nistari* variety has become a thing of the past in Tripura. Introduction of new races like $PM^{X}NB_4D_2$ almost doubled the productivity. Subsequently, other breeds like PM $^{X}CSR_2$, PM $^{X}FC_2$were introduced, which gave higher productivity. PM ^{X} SK_6SK_7) race was introduced later in 2011-12, which showed higher resistance capacity to diseases and had higher productivity of 40 to 45 Kg per 100 Dfls. Some features of the silkworm races reared are given in Table 6.3.

Table 6.3: Silk worm races reared

Race	*Period*	*Productivity*	*Disease Resistance Capacity*
Nistari	1976-1995	15-20 kg/100 Dfls	High
PMX NB4D2	1995-2004	30-35 kg/100 Dfls	Average
PMXCSR2	2004-2010	35-40 kg/100 Dfls	Susceptible to Disease
PMXFC2	2010-2011	35-40 kg/100 Dfls	High
PMX(SK_6XSK_7)	2011-2012	40-45kg/100 Dfls	Average

Out of various components of Sericulture, rearing of silk worm is the most critical in producing quality cocoon and giving adequate return to the rearers. In this context, introducing pure *bivoltine* and multi-voltine ^{X} bivoltine hybrid for improving the quality has been one of the major interventions. In sericulturally advanced states like Karnataka, there is resistance to introduction of bivoltine race in place of multi-voltine due to several factors. However, in new areas there was no such resistance from the rearers, as they had no choice but to go by the official recommendations.

Under TWSP, rearing of four crops a year was advocated under rain fed condition for maximising return to the rearers. The tentative time schedule followed is given in Table 6.4.

Introduction of the rearing of improved breed of silkworm under the project TWSP ensured improving the quality of silk yarn as well as return to the rearers. The high quality of silk yarn produced in the State was found suitable for weaving in handloom as well as power loom.

Table 6.4: Schedule of rearing during the year (season wise)

Crop No.	*Seasons*	*Silkworm race/breed*	*Time of brushing*
1	Feb-March.	BV Hybrid (CSR2 X CSR4)	2nd week of February
2	May-June.	(PMx CSR2), PM X (SK6 X SK7)	2nd week of May
3	Aug-Sept.	(PMx CSR2) PM X (SK6 X SK7)	2nd week of August
4	Oct-Nov.	(PMx CSR2) PM X (SK6 X SK7)	3rd week of October

Grainage for Production of Silkworm Egg

Besides mulberry leaves, Silkworm egg or seed, commonly known as Disease Free Laying (Dfl) is another basic input for the sericulture industry. Silkworm seeds are produced in 'Grainage' derived from the French word '*grana*' which means eggs. In Tripura, initially required Silkworm eggs used was supplied by the Central Silk Board (CSB) from its grainages. It was brought from Bangalore to Tripura by air and there were loss and damage due to transportation over long distance. Further, acclimatization was also a problem which was making the silkworm susceptible to disease and was a matter of concern. Setting up of a modern grainage was taken up with the financial assistance and technical support of CSB at Agartala, the state capital, for meeting the demand and also to overcome the risk of transportation over a long distance. The State grainage set up at a cost of Rs. 87.57 lakh was inaugurated in 2005. An account of silk worm eggs(Dfl) produced in the grainage is given in Table 6.5. The grainage was run departmentally. The average cost of production came to about Rs 400/- per 100 DFL during 2012-13.

Special rearing was taken up for production of seed cocoon with selected seed rearers near the departmental farms under close supervision of the trained officer and staff. The state Grainage Unit procured seed cocoons from the seed rearers, arranged its sorting, storing and took up production of Dfls as per requirements. This unit maintained different silk worm races

Table 6.5 : Production of Disease Free Layings in the State Grainage

Year	*Dfls (no.)*
2005-2006	14,000
2006-2007	25,000
2007-2008	42,000
2008-2009	24,000
2009-2010	26,000
2010-2011	33,000
2011-2012	45,000
2012-2013	67,000

with the help of Research Extension Centre, CSB and Mulberry Extension Centre, Champaknagar. About 30 per cent of the demand of Dfls is being met from the State Grainage at present and remaining is being procured from CSB. There are plans to strengthen the facility further for meeting the full requirements of the state and supply to the neighbouring states.

From the quantity of Dfls produced in the State Grainage Unit, given in Table 6.5, it appears that production of silkworm seed went up from 14,000 in 2005-06 to 42,000 in 2007-08. It appeared that there was a decline in the production during 2008-09 and 09-10, which again went up to 45,000 in 2011-12. However, there was no report of any disease, which indicates good management of the grainage.

Analysis of the production of cocoon per 100 Dfls, given in Table 6.6, indicated that it went up from 32.4 Kgs in 2005-06 to 42.5Kgs in 2011-12, which amount to increase by about 31per cent. Rearing of better breed of silkworm as well as higher efficiency in rearing appears to lead to this increase.

Table 6.6: Production of Cocoon per 100 Dfls

(in Kg.)

Year	*Race (CB)*	*Cocoon /100 Dfls*
2005-2006	PMXNB4D2	32.4
2006-2007	(PMXNB4D2)	33.1
2007-2008	(PMx CSR2)	35.7
2008-2009	(PMx CSR2)	37.2
2009-2010	(PMx CSR2)	37.85
2010-2011	(PMx CSR2)	39.8
2011-2012	PMx(SK6SK7)	42.5

Hatching percentage and other details of Dfls are given in Table 6.7. Hatching percentage was more than 90 per cent except for the period Sept-Nov., 2009. During this period, the number of late borne eggs had increased to high levels, which was corrected subsequently. The high hatching percentage and productivity per 100 Dfls indicated success of the grainage in a new area.

Table 6.7: Quality of silk worm egg

Crop Season	*Eggs/ Dfls (Nos.)*	*Dead Eggs/ Dfls (Nos.)*	*Unfertilized Eggs/ Dfls (Nos.)*	*Late borne Eggs/Dfls (Nos.)*	*Hatched Eggs/ Dfls (Nos.)*	*Hatching per cent*
Oct-Nov, 2007	492	3	9	20	460	93.49
May-Jun, 2009	450	3	9	18	420	93.33
Sept-Oct, 2009	442	7	3	37	395	89.36
Oct-Nov, 2010	432	6	4	5	417	96.52
Feb-Mar, 2011	436	4	8	15	409	93.80
Sep-Oct, 2012	497	3	5	9	480	96.57
Feb-Mar, 2013	503	4	4	9	520	96.73

Chowki Worm (Young Silkworm) Rearing

Chowki rearing involves scientific management of young silkworm larvae for 8 to 10 days up to the second instar. It plays an important role in ensuring success of the crop and getting higher yield of cocoon. Keeping in view its advantages over rearing from Dfls at the farmers' level, rearing of young silkworm (Chowki worm) was taken up in the Mulberry Extension Service Centre set up under the TWSP. Subsequently, the women groups were encouraged for taking up chowki worm rearing by contributing their labour under supervision of the Mulberry Rearers Sericulture Co-operative Society. One Chowki rearing centre was to be set up for each 50 acre of mulberry plantation. A package of assistance covering high yielding Mulberry Variety for raising of chowki mulberry garden, irrigation for mulberry garden, construction of chowki rearing house, disinfectants like *Chlorofect*, lime, Bleaching powder, detergent for disinfection of rearing house and appliances, *Labex*, *Sanjeveene* for bed disinfectants, tray dala, nets, as well as support for transportations of chowki worm were provided under the project.

Rearing of Chowki worm was done by a number of agencies, which can be grouped in to four broad categories namely by departmental units, by SHGs, by selected private rearers and individuals. Analysis of the trend of chowki rearing, given in Table 6.8, indicate that chowki rearing increased from 75,000 units in 2007-08 to 1,51,000 units in 2011-12. However, during the same period chowki rearing in the departmental units declined from 23,000 in 2007-08 to 10,000 in 2011-12. This trend appeared to be good as it indicates that the role of the department has been changing from that of a performer to a facilitator.

Table 6.8 : Chowki (young silkworm) rearing

Year	*Dept.(MEC,CRC)*	*SHG*	*Private Rearer*	*Individual*	*Total*
2007-08	23,000	14,000	9,000	29,000	75,000
2008-09	17,500	15,500	11,000	34,000	78,000
2009-10	16,000	17,000	10,000	38,000	81,000
2010-11	16,000	25,000	18,000	86,000	1,45,000
2011-12	10,000	30,000	20,000	91,000	1,51,000

Return to the rearers having mulberry plantation depends on number of rearing on crops of silkworm taken up in a year. Under rain fed, as many as four crops can be taken up in a year. However, the rearers take up number of crops based on yields of mulberry leaves and entrepreneurial ability of the rearers. The number of rearers taking up multiple crops in a year is given in Table 6.9. From the said data, it appears that initially in 2006-07 the highest number of rearers was rearing three crops in a year. The number of rearers rearing crops increased consistently in the subsequent years, being the highest in 2011-12. This was followed by number of rearers rearing three crops, two corps and one crop respectively.

Table 6.9: Cropping pattern of rearers

(no of rearers)

Crop pattern	*2006-07*	*2007-08*	*2008-09*	*2009-10*	*2010-11*	*2011-12*
4 crops	80	103	120	123	309	365
3 crops	102	109	137	165	251	320
2 crops	74	91	125	142	180	201
1 crop	45	15	56	73	107	125

Production of Cocoon

Cocoon produced are normally graded in to three grades based on silk content as determined from the number of cocoons per kg. 700 or less number of cocoons per Kg is taken as the Grade A representing the best quality, whereas 701 to 850 number of Cocoons per kg is taken as the next grade, Grade B, and more than 850 number of cocoons per kg as the third category, Grade C.

From the data given in Table 6.10, it appears that during 2008-09 to 2012-13 the five years period, total production of cocoon increased by three fold. Production of Grade A also increased substantially. This indicated the success of sericulture in the new area. Average production of cocoon is given in Table 6.11.

Table 6.10: Grade Wise Production of Cocoons in Tripura

(in Kgs)

Year	*A Grade*	*B Grade*	*C Grade*	*Total*	*Percentage of A grade cocoon*
2006-07	10,657.6	2752.55	2365.4	15,775.55	67.5
2007-08	13,719.1	2466.25	2745.3	18,930.65	72.4
2008-09	14,585	4045	1349.4	19,979.4	73
2009-10	18,557	5326	1910.9	25,793.9	71.9
2010-11	46,026	12618	2536	61,180	75.2
2011-12	48,336	13057	2315	63,708	75.8
2012-13	49,980	13858	1670	65,508	76.3

Table 6.11: Average Cocoon production

Year	*Average production (in Kg /ha/yr)*
2009-2010	150
2010-2010	260
2011-2012	290
2012-2013	310

Average annual growth of cocoon production has been analysed cluster wise and given in Table 6.12. It appears that the older clusters like Champaknagar, Santirbazar showed relatively good growth, whereas the position in respect of the new clusters like Hurua and Udaipur was not up to the mark.

These appeared to be due to inadequate infrastructure and lesser technical person available in the relatively new clusters. The trend of growth appeared to be same for total cocoon as Grade 'A' cocoon.

Table 6.12: Average Annual Growth Rates (AAGR) of Cocoon Production

Clusters	*Total*	*Grade A*
Champaknagar	25.62	28.47
Bishramganj	14.21	24.30
Halahali	14.14	18.38
Hurua	9.78	11.23
Kanchanpur	23.56	32.02
Udaipur	10.27	9.80
Santir Bazar	27.08	25.49
Tripura (average)	18.22	16.56

Note: AAGR calculated by fitting the line y=b*m^x, where, AAGR=(b-1)*100

Rate of Purchase of Cocoons

Cocoon produced by the rearers is purchased by the primary society as per the rate notified. The rate was revised from time to time keeping in view the price of silk. The price was increased in 2009-10 and subsequently in 2010-11.An account of the rate is given in the Table 6.13.

In Tripura, the Mulberry Rearers Cooperative Society (MRCS) organized at cluster level provides the institutional support for the purchase of cocoon at the above rate fixed for the entire State. The freshly harvested cocoon is stifled and stored in the godown of the society. Initially, electric hot air drier

Table 6.13: Rate of cocoon for purchase from the rearers

(in Rs/kg)

	Cocoons grade	*2002-03*	*2009-10*	*2010-11*	*2012-13*
1	A grade (nos. cocoons per Kg up to 700 nos)	Rs.90/-	Rs.120/-	Rs.200/-	Rs. 220/-
2	B grade (nos. cocoons per Kg up to 700-850 nos.)	Rs.70 /-	Rs. 80 /-	Rs. 150/-	Rs. 160/-
3	C grade (nos. cocoons per Kg up to above 850 nos.)	Rs.30/-	Rs. 40/-	Rs. 100/-	Rs. 85/-

installed in the departmental farms was used for stifling green cocoon. However, frequent disruptions in electric power supply posed constraint in using these machines. With a view to overcome this, Ushmakoti (drying chamber) operated by hot air circulation made up by burning mulberry twigs as is done in China was set up in each cooperative society. This has been found to be effective in stifling cocoon and reducing loss due to emergence of moth. The loss incurred due to improper stifling has been reduced thereafter.

Return to Rearers

An account of income of the rearers from sale of Cocoon has been given in Table 6.14 separately for each cluster. It appears that Santir Bazar has topped the list among the clusters in production, while same was lowest for Kanchanpur. In terms of average income per beneficiary, it is observed that Santir Bazar had the maximum average income per beneficiary in 2008-09. This was replaced by Champaknagar since 2009-10. Bisramganj and Champaknagar clusters had shown steady growth. The rearers of Takarjala also had a good year in 2010-11 and 2011-12, whereas the situation at Kanchanpur was not impressive.

Table 6.14: Income of rearers from sale of cocoon

Cluster	*Value of Cocoons (Rs. Lakh)*				*Average Income per Beneficiary (in Rs.)*			
	2008-09	*2009-10*	*2010-11*	*2011-12*	*2008-09*	*2009-10*	*2010-11*	*2011-12*
Champaknagar	17.15	18.30	21.15	28.35	6100	6700	7033	12400
Bishramganj	16.97	17.85	21.35	25.78	6106	6750	6800	11800
Takarjala	2.53	2.80	4.55	1.43	4100	4745	6790	10200
Udaipur	5.95	6.65	12.86	10.73	5200	5495	6153	12300
Amarpur	7.91	6.12	10.32	14.22	6500	5619	6370	12700
Santir Bazar	16.98	18.37	23.27	23.59	6505	6200	6610	12300
Halahali	5.42	5.95	9.10	6.55	4500	5130	5582	10200
Panisagar	5.60	6.65	8.92	6.03	4600	5162	4309	9100
Hurua	4.72	5.075	10.32	6.09	4495	4500	5134	9400
Kanchanpur	2.71	0.35	0.52	1.41	1571	1400	2080	6700
Total	85.94	88.12	122.36	124.18	4967.7	5170	5686	10710

Bulk of the income on sericulture comes from sale of cocoon produced by rearing silkworm. The rearers also get some income by sale of cuttings from the mulberry plantation as well as from sale of saplings raised in decentralized people's nursery. With a view to assess the return to rearers from production of cocoon, an analysis was undertaken in respect of the achievement in ten clusters where sericulture was undertaken. From the performance of rearers in terms of sale of cocoon, it appears that in respect of six clusters the income has remained more or less steady, around Rs 6,000/- per year during the period from 2008-09 to 2010-11. The level of income increased substantially during 2011-12, following increase in the rate at which cocoon was purchased from the rearers by the society. Out of ten clusters, performance of the rearers was reasonably better in six clusters, namely Champaknagar, Bishramganj, Takarjala, Udaipur, Amarpur and Santir Bazar. However, the remaining four clusters at Halahali, Panisagar, Hurua and Kanchanpur, income of the rearers has been much lower. The above analysis prove the point that sericulture activity in new area should be taken up only in a few selected clusters where provision of required forward and backward linkages and extension staff in adequate number have been ensured. Undertaking this activity at several places without supporting infrastructure and extension staff may yield result in the short run but fail to sustain on a long term basis. It would be appropriate to avoid such experiment.

With a view to analyse the performance of rearers belonging to different categories, the performance of the rearers in respect of one selected cluster namely Bishramganj Sericulture Cluster was taken up for analysis. The performance of rearers belonging to Scheduled Tribe, Scheduled Caste, Minority, OBC and General categories was analysed separately for each group in terms of productivity of cocoon and income from sale of cocoon. From the analysis done over a period of five years, 2006-07 to 2010-11, it appeared that beneficiary belonging to Minority category performed relatively better, followed by General category and Scheduled Tribe category and performance of OBC and Scheduled Caste category remained somewhat low. As the clusters are relatively small, it may not be proper to draw any conclusion from this analysis. It may be presumed that the performance is neutral to caste and group considerations. All categories of rearers performed equally well.

Reeling of Cocoon

The process of value addition in Sericulture occurs over various stages. The cocoons are reeled to yarn, twisted and then weaved to the marketable form, mostly sarees. The process of marketing also involves certain economic implications as a result, value addition occurs at that stage also. The major problems relate to non-availability of skilled reelers and low margin. This has been overcome by taking up reeling in two parts— (i) by advanced charkha at the Society level and (ii) with multi end reeling machines at the State level. Relatively poor grade of cocoon was reeled by using advance charkha for development of skills. Superior grade of cocoon were sent to the State level reeling unit (filature) at Badharghat.

Details of the cocoon processed in the Badharghat reeling units, given in Table 6.15. Reeling machines purchased earlier were lying idle for a long time. These machines were made operational by getting skilled workers from Karnataka, who installed the machine and imparted training to local youth on turn-key basis. This has paid rich dividend by introducing reeling in the non-traditional area. Cost of production of 1 kg of raw silk increased by Rs. 560/kg due to increase in raw material cost as well as labour. However, the society is hopeful of getting about Rs. 32/kg from the reeling stage which can be used for maintenance and depreciation. The pricing of the reeled silk is

Table 6.15: Economics of Reeling

	2006-07	*2007-08*	*2008-09*	*2009-10*	*2010-11*	*2011-12*
Purchase of Cocoons (Kgs)						
A-GRADE	9011	10737	9543	8138	11096	17761
B-GRADE	1874	2223	3251	2823	3426	4127
TOTAL	10885	12960	12794	10961	14522	21888
Yarn Reeled (in Kg)	1020	964	1262	1076	1312	1970
Yarn Sold (in Kg)	822	964	1262	977	384	—
Value of Yarn (in INR)	1028322	1113420	1457610	1128435	2361600	3743000

partly subsidised. A comparative account of costing of reeled silk yarn during 2002 and 2013 is given in Table 6.16.

Table 6.16: Economics of reeling silk yarn

(Rs/Kg)

Component	*Year*	
	2002	2012
Raw materials (8 kg cocoons)	880	1400
Wages (Reelers &Twister)	145	300
Fuel (Gas & electricity)	100	50
Electricity for stifling & other purposes	15	30
Gross production cost	1140	1780
Income from by-products (floss)	00	112
Net production cost	1140	1668

Twisting of Yarn

Twisting of silk yarn is the second stage of processing. It is necessary for giving additional strength and holding capacity to the silk yarn, before taking up weaving. The state had procured twisting machines, but same were lying unutilised in absence of skilled persons. Special efforts were made for making these twisting machines operational by getting expert twisters from Karnataka in 1995 who gave on-the-machine training to youth from the local area in twisting. Subsequently, the twisting machines are functioning properly at the Badharghat Reeling Unit. The twisted yarn is sent to the Power-loom unit, Indranagar for use in weaving. The financial aspect of twisting is given in Table 6.17.

Table 6.17: Economics of twisting silk yarn

(value in Rs. quantity in Kg)

Component	*YEAR*		
	2002	2010	2012
Cost of raw Silk/Kg.	1140.00	1140.00	1400.00
Wages	180.00	500.00	500.00
Total	1320.00	1640.00	1900.00

Weaving of Silk Fabrics

Weaving of twisted and plain silk yarn in hand operated/machine operated loom represents the last stage of sericulture. In Tripura, silk weaving was initially taken up by the handloom clusters developed under the women sericulture projects. However, in due course there has been decline in the number of commercial handlooms in Tripura. This has led to problem in silk yarn converting to fabrics and stock of silk yarn piled up. With a view to overcome this weaving by silk power loom was introduced in the state. Grey silk fabrics produced were sent to Kolkata for printing and processing. However, of late, screen printing, hand printing and embroidery work were taken up on silk fabrics in the state itself leading to additional value of fabrics and generation of employment facility for the youth.

Power loom was installed in departmental units. Initially, four numbers of looms were set up in Indranagar in 2010. Subsequently, eight more looms were installed at Badharghat in 2012. The approximate cost of one unit including transportation charge comes to Rs 1.95 lakh.

The estimated cost of printing a silk saree in these centres comes to aboutRs. 1450/- per piece, details of which is given in Table 6.18. This estimate is based on the assumption that one silk saree weighing about 400 grams and silk yarn as the raw material has the largest share. A weaver weaves about 2 (two) sarees in 3 (three) days, i.e. 20 sarees in the month. Weaver's wage comes to Rs. 7000/- per month on an average.

Table 6.18: Estimated Cost of Printing

Component	*Rs./Unit*
Raw material	680
Weaving	350
Printing	300
Weaving wastage.	20
Transportation	50
Misc.	50
Total	1450

Account of value addition in processing silk yarn done in the state is given in Table 6.19 and Table 6.20. The production increased both in physical and financial terms. The production of raw silk sarees increased from 520 numbers with a cost of Rs. 3.16 lakh in 2007-08 to 2225 pieces valuing Rs. 21.80 lakh in 2010-11. The increase in production of saree also led to increase in wage of the weavers. It has been found that the earning of the weavers increased by about nine times during the period 2007-08 to 2010-11.

Table 6.19: Value Addition in Processing

Stage	2007-2008	2008-2009	2009-2010	2010-2011
1. Silk yarn (in kg)	128	2165	2285	840
2. Silk Sari-Raw (in pieces)	520	2165	2285	2225
3. Silk Sari-sent for printing (in pieces)	520	2121	2245	1703
4. Silk Sari-received from printing (in pieces)	433	2121	2245	425
5. Sale of Silk sari (in pieces)	345	1736	1824	1784
6. Stock in hand (in pieces)	88	429	421	500
7. Weavers paid for (no)	520	2165	2285	2225
8. Weavers engaged (no)	30	40	40	40

Table 6.20: Value Addition in Marketing

(in Rupees)

	Stage	2007-08	2008-09	2009-10	2010-11	2011-12
1	Silk yarn	229320	1091160	1151640	1112500	3743000
2	Silk Sari -Raw	316000	1875337	2010800	2180500	2565500
3	Silk Sari- received from printing	436103	2435487	2685460	416500	439800
4	Sale of Silk sari	339303	2398849	2571840	2854400	3665000
5	Payment to Weavers	87315	624004	685500	778750	8063000
6	Printing cost	138876	377653	508488	553475	652375
7	Packing, transportation etc.	66650	54629	132087	45740	57200
8	Surplus	54880	352640	383442	367509	2106344

Innovative Measures Adopted in Tripura

Synergy in Promotion of Sericulture with Implementation of MGNREGA

The National Rural Employment Guarantee Act, promulgated

by the Parliament on 23rd August 2005, has been landmark legislation in the country. This legislation has made provision for employment for 100 days in a year mandatory in rural area all over the country. Besides, provision of employment, the Act envisages empowering the local communities and strengthening the livelihood resource base of the rural poor.

In Tripura, the scheme was introduced in Dhalai District during 2006. Subsequently, South Tripura and West Tripura districts were covered during 2007-08 and North Tripura district has been included during 2008-09 along with other districts in the country. In Tripura, there are 6,11,732 numbers of job card holders under the programme. Keeping in view the advantages of Mulberry Sericulture in providing a steady source of income to the women belonging to the Scheduled Tribe, fund available under the MGNREGA was dovetailed for taking up Mulberry sericulture.

Implementation of MGNREGA with the active involvement of the 3 tier Panchayati Raj structure marks a special feature of Tripura. An account of fund received and spent during 2007-08 to 2011-12 is given in Table 6.21. It is noteworthy that Tripura got award for best performance under MGNREGA in the country during 2010-11 and 2011-12. Resource available under MGNREGA was availed for undertaking projects relating to fishery, sericulture, forests, agriculture, irrigation, horticulture, bamboo and tea plantation, etc.

Table 6.21: Fund received and spent during 2007-08 to 2011-12

(amount in Lakh Rs)

Financial year	*Fund received*	*Expenditure*	*% of expenditure*
2007-08	21850.39	20860.33	95%
2008-09	51400.27	49047.92	95%
2009-10	74288.80	72615.33	98%
2010-11	63276.16	62549.89	99%
2011-12	100870.22	94559.93	94%

Source: Achievement Report (2007-12), Rural Development Department, Government of Tripura

With a view to ascertain income generated from mulberry sericulture vis-à-vis MGNREGA, available data was examined by taking 30 per cent rearers in one of the sericulture clusters selected at random, details of which are given in Table 6.22.

From the analysis, it is evident that Sericulture activities from the number of dfl consumed increased from 80 in 2008-09 to 94 in 2009-10 and to 150 in 2011-12. This means that the productivity level increased continuously during the said three years. However, income from Sericulture went up substantially during 2009-10, which was partly due to production of more cocoons and partly due to upward revision of rate at which cocoon was being purchased from the rearers by the mulberry cooperative society during 2009-10.

Table 6.22: Income from sericulture, MGNREGA, and other sources

Activity	*2008-09*	*2009-10*	*2010-11*	*2011-12*
Sericulture				
Dfls Consumed in nos.	80	94	105	150
Cocoon produced in Kg.	23	36	41	42.5
Income from sale of Cocoons in Rs.	2212	7150	8059	11500
Income from sale of Saplings in Rs.	1283	1107	1210	2500
MGNREGA				
Man-days generated in days	53	68	64	85
Wage generated in Rs.	4336	6510	6486	11030
Income from other sources in Rs.	3448	3997	5001	6067
Comparison				
Total income (Sericulture & MGNREGA)	11279	18764	20755	31097
Sericulture income as % of total income	31	44	45	45

Infrastructure for the Development of Sericulture available at Tripura

Keeping in view the importance of infrastructure in development of sericulture on a sustainable basis, specific stress was laid for development of required infrastructure for supporting sericulture. Resources available under various ongoing central and state sector schemes were pooled for the purpose. An account of the infrastructure created for supporting sericulture is given in Table 6.23.

Growth of Sericulture

The Sericulture sector in Tripura continued to get support after the project period. The state government took steps to promote

Table 6.23: Infrastructure created for supporting sericulture

Sl No	*Particulars*	*No of units*
1	Sericulture training Institute	1 no.
2	Silkworm Seed production centre (Grainage)	1 no.
3	Mulberry Sericulture Cluster	18+1*
4	Mulberry Rearers cum Reelers cooperative society Ltd.	16 nos.
5	Research Extension centre	1*
6	Area under Mulberry plantation (in acre)	4375+2*
7	Sericulture beneficiaries	9012
8	Reeling Units	4 nos.
9	Twisting units	4 nos.
10	Silk processing unit	1 no.
11	Silk Power loom unit	2 nos.
12	Silk marketing centre	2 nos,

(* in central sector)

this activity with provision of technical and financial support. It was found that the number of rearers increased so also the production, income and the number of people involved in these activities. It appears that the total number of beneficiaries increased substantially. It appears that interest on sericulture varied from cluster to cluster and from year to year. The interest of the field staff and officer and the local leadership played an important role in this.

The year wise progressive achievement of number of beneficiaries, area under mulberry plantation, cocoon production, etc. is given in Table 6.24.

Table 6.24: Achievements of Mulberry sericulture Tripura

Parameters	*2007-08*	*2008-09*	*2009-10*	*2010-11*	*2011-12*	*2012-13*
1. Beneficiaries (no)	2531	873	1208	1250	1323	1827
2. Area under mulberry plantation (acre)	1265.5	436.5	604.0	625.0	661.5	717.0
3. Cocoon produced (MT)	16.55	17.50	19.26	29.43	31.39	55.53

The number of beneficiaries taking up sericulture went up steadily. About 9,012 new families were added in this activity, which amounted to about three fold increase. Similarly, as the number of beneficiaries is increasing every year, the plantation area also increased in almost same proportion and the plantation area increased by 3.4 fold between 2012-13 and 2007-08. The beneficiaries and mulberry plantation coverage area increased and cocoon production increased by more than three times.

Lessons Learnt

Experiences of promotion of mulberry sericulture on commercial scale in Tripura, spanning over the last nearly three decades, since 1995 when women sericulture project was launched, provide several important lessons on development of mulberry sericulture and production of Silk in new and non-traditional areas, some of which include –

(i) Establishment of proper linkages covering plantation, rearing of silkworm, reeling, twisting and weaving are of critical importance for making the operation sustainable. Success of the practice in Tripura is attributed in particular to the provision of two major forward and backward linkages (i) Silkworm Grainage for production and supply of disease free laying (DFL) produced under local conditions which is more hardy and disease resistant and (ii) arrangement for marketing of cocoon covering reeling and twisting of silk yarn along with forward linkage of weaving silk fabrics on handloom/power loom, printing and processing it as per the market demand.

(ii) Adoption of appropriate technology in plantation of mulberry variety, rearing of silkworm races, stifling of cocoon, reeling and twisting are important features, which have contributed positively to success of the practice.

(iii) Generation of awareness about the benefits of sericulture, capacity building by organizing decentralized training and provision of technical guidance for raising multiple crops helped in increasing production, productivity and return from sericulture.

However, there are a lot of things to be done for increasing productivity, quality and return to the level of the traditional states. This would require provision of technical guidance both for plantation and rearing of silkworm. Provision of irrigation sources will increase production of mulberry-leaves and enable taking up more silkworm for rearing. The productivity level in terms of cocoon produced from 100 DFL in the State is close to average production level of the traditional States. This indicates that the transfer of technology is more or less complete and the new rearers have been able to adopt new technologies. As such, provision of irrigation sources and other inputs for production of more leaves will enhance the quantity of silkworm to be reared and production of cocoon from unit area. On the other hand, the arrangement for weaving silk fabrics is required to be stepped up as this is a critical thing in taking the product to the market.

Most of the farmers in the State are small and marginal. Even though attempts were made for encouraging bigger farmers to take up sericulture, it has not met with significant success. With the implementation of MGNREGA, the wage structure for unskilled manual labour has gone up substantially. Further, sericulture has to compete with vegetable cultivation which gives higher and assured return and more is secure as compared to silkworms rearing which is more susceptible to disease. This makes sericulture more risky than that of vegetable cultivation.

In the State, attempts have been made for dovetailing MGNREGA wage for promotion of sericulture in case of Scheduled Tribe, Scheduled Caste, small and marginal farmers particularly in maintaining the plantation, construction of rearing house and taking up rearing. This is a novel method of implementation of MGNREGA, which make the expenditure more productive, provides employment and income to the target group on a sustainable basis.

In Tripura, about sixty per cent of the total land is recorded as 'forest'. This is mostly upland (Tilla) with undulating topography. Right to use of 1.7 lakh hectare of forest land has been given to 1.1 lakh forest dwellers in the State. As there is no restriction on raising mulberry tree in the forest area, tree-type mulberry can be raised in such forest land and the leaves so produced can be used for promotion of sericulture.

One of the noticeable features in the success of the programme lies in providing the linkages. This has been possible only by restricting the practice to a few geographical clusters. At times, there are temptation for expanding sericulture to new areas without adequate support and infrastructure both physical and administrative. Notwithstanding the amount spent, such measures have always been ending in failure which should be avoided at all cost. Due care should be exercised while planning for expansion of sericulture to new areas. It should be at best restricted to the Gram Panchayat adjacent to existing clusters and new areas at a distance from clusters should be avoided.

The fact that sericulture has been taken up in the State mainly by women and that too belonging to Scheduled Tribes is another important factor. It has helped in empowering disadvantaged segments like Scheduled Tribe population in general and bringing out better gender equality by empowering the women in particular.

The experience of Tripura throws light on the lines mulberry sericulture can be developed in new and non-traditional area on a sustainable basis. This will be of use to other States interested for taking up mulberry sericulture for generation of sustained livelihood particularly the tribal and small and marginal farmers.

The Sericulture sector as analysed in the previous sections allows several inferences. First and foremost is the fact that the sector provides indication of sustainable growth and expansion in all spheres like—production, value addition and employment. However, this process of growth and development could be possible only due to effective intervention of the government in the recent years, in issues like marketing and technical assistance. The private traders, earlier, had to go for distress selling due to low demand within the state, but the government has initiated the process of establishing four new power looms so that the entire raw silk of the state can be fruitfully utilised. It was also seen that in Tripura, since 1994-95 a large number of weavers of Handloom sectors started grey silk saree production on pit loom. But due to low wages and isolation by the marketing agency the weavers diverted into other production. In the year 2007-08, a challenge was taken by the Directorate for revivification for production and marketing of the silk products through Primary Weavers Cooperative Societies. Initially, 5 (five) Primary Weavers Cooperative Societies was selected for production of silk item by increasing wages as well as marketing assurance. Presently, there are 50 (fifty) weavers producing silk fabric in the state.

Subsequently, the tribal weavers also started producing value added silk fabric through loin loom with decorative tribal motive. To facilitate easier supply and sales of quality products to the customer, a modern sales outlet namely – 'TRIPURA SILK SHOW ROOM' has started for marketing exclusive silk fabric since 2007-08. The show room is registered under SMOI for 'SILK MARK' to ensure quality for the consumers. The First Lady of Tripura released the 'Silk Mark Level' in the state in 2008-09. Moreover, to cater to the growing demand for new designs and varieties, a printing unit was established at Indranagar, Agartala where facilities like silk processing, printing and finishing of silk fabrics are available. The process of value addition within the sector has, therefore, received a tremendous fillip at these public sector initiatives.

On the other hand, at the farm level also, impressive progresses are visible. The cluster based growth strategy has resulted in sustainable development, with useful economic implications for the beneficiaries. The total number of beneficiaries for cocoon production has increased by almost 250 per cent, while output has increased by more than 80 per cent. However, progress has not been uniform in all clusters. Champaknagar, Bishramganj, Santir Bazar the well-endowed cluster while Panisagar falling behind. Moreover, data reveals that production of Mulberry, Cocoon and Raw Silk have increased over the years and the fact becomes more impressive following the evidence of dominance of Grade 'A' Cocoon. The economics of the sector had a boost with the enlargement of the revolving fund following increasing allocations from the Central Silk Board and the MGNREGA scheme. Nevertheless, the progress at the farm level could also be obtained due to progressive and necessary intervention of the government particularly for grainage and silkworm rearing. The Directorate of Handloom, Handicrafts and Sericulture took special care for facilitating beneficiaries with the production of seed cocoons and supply of Dfls to commercial rearers with fruitful results as is observed in the increasing income profile of the beneficiaries across all clusters. However, the case study on Bishramganj cluster reveals that all communities are not able to reap in the benefit evenly. Cocoon growers from the General, Religious Minorities and Scheduled Tribes community seem to harvest better crop and income while growers belonging to the

Scheduled Castes (SC) and Other Backward Classes (OBC) are found to be lagging behind.

Factors, those Contributed to Stabilising Sericulture in Tripura

One of the basic requirements for stabilizing sericulture in a new area relates to provision of forward and backward linkages. In this respect, the forward linkage in terms of arrangement for procurement, stifling, storing, reeling of cocoon and twisting yarn are of major concern. This need is more conspicuous as mulberry cocoon is perishable and in new areas the rearers have no mechanism for selling the cocoon of their own. Due to lack of volume and relatively less private growers it would be futile to expect the private sector for taking up this activity at the initial stage. This is in fact the main reason limiting promotion of sericulture on a commercial scale in new areas. In Tripura, purchase of stifling and storing of cocoon was taken up through society set up at the Block level catering to the need of 2–3 clusters in the area. Necessary infrastructure for purchase, stifling and storing of cocoon was developed under the primary rearers' cooperative society. This experiment appeared to have yielded good result.

Reeling and Twisting of Silk Yarn

The second component relates to reeling of cocoon. Reeled yarn is twisted before it is used for weaving silk fabrics. The reeling of cocoon is a highly skilled job and the margin available is much low. In case of twisting, the margin is relatively higher. Apart from margin, maintaining quality of the product and reducing wastage are other aspects which need close attention. Like other new States, reeling and twisting machines were purchased and installed in Tripura. However, those were not put into use for a very long time. Non-availability of skilled manpower was the main factor behind this. Efforts made by Govt. of Tripura in getting skilled, reelers from Karnataka, for giving hands on training to the local reelers is an excellent initiative. It has been found that as a result of this effort taken up under turnkey basis, the reeling machines could be put into use and over a period of time, the local youth acquired required skill in reeling and produced best quality bivoltine silk. The use of relatively inferior (B) grade cocoon for reeling in advanced charkha at the primary society level is yet another innovative step. This helped in converting cocoon to yarn, providing wage employment and reducing the cost on account of transportation to the reeling unit. Steps were taken for twisting of yarn by using skilled twister who was brought from Karnataka. The quality of twisted yarn has been found to be very good. This is yet another component of critical importance in stabilizing sericulture in a new area like Tripura.

The above experience of setting up the arrangement for reeling and twisting on a decentralized manner and doing it through skilled reeling and twisters from Karnataka for imparting skill while working on the machine is an excellent initiative. This is required to be replicated for sustaining sericulture on a long term basis.

Weaving Silk Fabrics

The third component relates to converting silk yarn to fabrics. Unlike traditional handloom states, Tripura has a relatively low base of handloom weavers. The fabrics woven by tribal families in loin loom are of relatively narrow width. Number of active commercial frame looms is limited. Further, this number has been declining on account of various reasons. As a result, it was found that bulk of the silk produced in the State was remaining unsold as there is nothing much demand in the local market. It was difficult to sell such low volume outside the state. With a view to overcome this, weaving silk fabrics with power loom was introduced in Tripura. The silk yarn available was converted to plain fabrics in the power loom. Said fabric was sent for printing and processing to Kolkata and sold back in the State.

Silk of Tripura has got market and has been appreciated in the market because of its quality and being available at attractive prices.

Silkworm Seed Production Centre (Grainage)

The other important component for stabilizing sericulture in a new area relates to backward linkage. The basic input in case of sericulture relates to mulberry cutting of high yielding variety and silkworm egg for taking up rearing. Improving quality of mulberry has been relatively easy, which is done by getting mulberry cutting of better quality from outside and converting it to mulberry plantation either by direct plantation with cuttings or by raising sapling, which is later transplanted in the field. In Tripura, the high yielding variety of mulberry was introduced at regular intervals based on the internal resource and the support extended by the Central Silk Board.

The supply of silkworm egg is another critical component in sustaining sericulture. Initially, Tripura was dependent upon the Central Silk Board for getting silkworm egg. It was important to set up a grainage for production of silkworm egg. It is encouraging that the grainage started in Tripura has been functioning well and producing good quality silkworm egg. Tripura has taken up rearing bivoltine silkworm, which has distinct advantage over the traditional States where the multivoltine and hybrid bivoltine verities are being tried. Setting up of the grainage along with supporting infrastructure like seed farm and seed rearers are the other innovations, which have far reaching effect on success on sericulture in the State. It has been possible because of the hard and sincere efforts made by the staff and officers of the Sericulture Department.

In brief, the experience of Tripura indicates that it is possible to introduce mulberry sericulture in new area by taking up various activities in a planned and coordinated manner and giving support over a longer period of time. This experience can be availed by other States keen for taking up sericulture. It would be worthwhile to replicate the success story by other States rather than attempting to reinvent the wheel. It appears that the Sericulture sector in Tripura is posed to grow in the future and contributing to the state economic development in general and making the growth socially inclusive in particular. However on order to remain competitive in the globalised economy, updating the level of technology and encouraging private sector participation need to be give special and continuing attention for making the growth sustainable.

Way Forward

Mulberry sericulture has several advantages such as needing less capital investment, being employment intensive and particularly useful to women who get the employment indoor as well as good market potential. Sericulture acts as a medium for transfer of resources from the rich, who purchases silk fabrics, to the poor, who produces silk cocoon and yarn. Mulberry sericulture is a traditional practice in Karnataka, Tamil Nadu, Andhra Pradesh and parts of Jammu & Kashmir and Bengal, it has been going on for centuries. With a view to extend its benefits to other states, commonly known as the non-traditional areas, number of plans and programmes has been implemented for last several decades. However, there has not been much success in extension of mulberry sericulture in new areas, where the achievement has remained rather limited, notwithstanding the public resources spent.

In this background, the experience of Tripura in introducing mulberry sericulture on a sustainable basis has been quite interesting. Success of mulberry sericulture in Tripura is attributed to a number of factors, as indicated below.

(i) Promotion of mulberry sericulture was kept confined to few selected geographical clusters and beneficiaries were selected in these clusters.

(ii) Officer and field staff was posted and trained for provision of required technical guidance.

(iii) Required assistance was provided to the interested villagers for taking up mulberry plantation, construction of rearing house as well as getting rearing equipment by dovetailing fund under departmental schemes as well as MGNREGA.

(iv) The rearers were organized into primary mulberry rearers' cooperative society at the cluster level. Required forward and backward linkages for promotion of sericulture were extended through the primary society. One departmental technical officer was posted as the managing director of the society, which ensured proper coordination between the department and the cooperative of the rearers.

(v) Arrangement for collection, stifling, storing and reeling of cocoon was taken up in a coordinated manner. Skilled reelers and twisters were brought from Karnataka for running the existing units in a viable manner.

(vi) The silkworm grainage was started for production of silkworm egg under local condition for meeting the requirement of the state.

(vii) The silk yarn was taken up for twisting and weaving. As the handloom base was not adequate, silk power loom was introduced for weaving the silk yarn in to fabrics.

(viii) Last, but not the least, important factor relates to promotion of sericulture under a mission mode through the Tripura Women Sericulture Project in 1995. Under the project, action was taken for building and maintaining critical infrastructure for marketing and processing of cocoon, production of silkworm egg as well as extension. On completion of the project, required assistance under the departmental scheme was provided for consolidating and continuing the achievement.

Conclusion

Keeping in view the multiple advantages of mulberry sericulture in providing employment in rural areas as well as production of silk, several attempts have been made to promote mulberry sericulture outside the so-called traditional areas. However, due to various reasons, such attempts have met with limited success. This has been due to the technicalities involved in mulberry sericulture, inadequate infrastructure, marketing facilities, supervision etc. The experience of Tripura is based on implementation of a women sericulture project covering generation of awareness, organizing primary producers into groups around geographical clusters, providing required assets and technology along with marketing linkages. Setting up of grainage for silkworm seed production and arrangement for marketing covering reeling, twisting, weaving of silk fabrics established a chain from "Soil to Silk". The set of photographs Fig. 6.1 to 6.30 indicates the women beneficiaries involved at various stages of the industry. It has resulted in the primary producers getting sizable income and production of quality bivoltine silk. Dovetailing of fund under MGNREGA for promotion of mulberry sericulture which has led to creation of productive assets along with provision of employment marked yet another interesting and innovative aspect of this work. It is beyond doubt that the experience of Tripura in promotion of mulberry sericulture producing better quality silk will show the way to expansion of sericulture in the new and non-traditional areas of the country in the days to come.

Fig. 6.1: Women rearers meeting in the MRCS

Fig. 6.2: Land preparation

Fig. 6.3: Land preparation

Fig. 6.4: Mulberry plantation irrigation

Fig. 6.5: Deweeding mulberry fields

Fig. 6.6: Mulberry leaf plucking

Fig. 6.7: Disinfection of rearing house

Fig. 6.8: Disinfection of rearing trays

Fig. 6.9: Grainage inauguration

Fig. 6.10: Silkworm egg preparation in grainage

Fig. 6.11: Preparing mulberry leaves for feeding silkworm

Fig. 6.12: Feeding silkworm

Fig. 6.13: Application of Resham Kit Aushadh

Fig. 6.14: Mounting of silkworm

Fig. 6.15(a): Cocoon Harvesting

Fig. 6.15(b): Cocoon harvesting

Fig. 6.16: Carrying of cocoon

Fig. 6.17: Rearers with produce at MRCS

Fig. 6.18: Purchase of cocoon in MRCS

Fig. 6.19: Receiving payment for the cocoon

Fig. 6.20: Cocoon stiffling in hotair chamber

Fig. 6.21: Usma Koti Stiffling chamber

Fig. 6.22: Silk Cocoon cooking

Fig. 6.23: Reeling of cocoon

Fig. 6.24: Silk yarn twisting

Fig. 6.25: Silk yarn packaging

Fig. 6.26: Silk weaving by handloom

Fig. 6.27: Silk weaving on powerloom

Fig. 6.28: Processing silk fabrics

Fig. 6.29: Silk Fabrics block printing

Fig. 6.30: Silk fabrics for sale

7

Future of Sericulture Industry

It has been the endeavor of Nations all over the world to usher into an era of development with peace and prosperity for the citizens. Development basically relates to the process of enlarging people's choice. Human choice covers a wide range, the most critical of which include living a long and healthy life, getting educated and having access to resources needed for decent standards of living. Additional choices include political freedom, guaranteed human rights and personal self-respect. Economic development usually leads to increase in availability of financial resources, which in turn expands the material need of individuals. This results in higher intake and greater pressure on the available natural resources, which is limited. As a result, making the development process sustainable over a long term has emerged as one of the major challenges. This in turn has laid greater thrust on sustainable development, which requires a pattern of resource use for meeting the human needs while preserving the environment. It is a pattern[1] which meets the needs of the present without compromising the ability of future generations to meet their own needs. Sustainable development ties together concern for the carrying capacity of natural systems with the social challenges facing humanity.

Sustainable development requires paying attention to issues falling in two broad categories namely social and environmental. Issues relating to socio-economic deprivation are perceived in the short run. As a result of technological innovation in the electronic media, information has been moving across the world virtually in no time and with little cost. This has led to increase in the expectation of the youth at a fast rate like at the speed of light. Even though the opportunities have been increasing, it has been at a slower pace, like at the speed of sound. The unmet expectations have been widening and often leading to frustration, which is giving rise to anger and it turning to violence. Active civil society and non-Government organizations along with the liberated spirit of the youth have added new dimension to this. This has been manifesting in several ugly and undesirable incidents at periodic intervals. As the rich-poor divide increases, such incidents are becoming more frequent and at times turning violent resulting in loss of property and life and retarding the pace of economic development. On the other hand, environmental issues have impact, which is felt on a longer term. Consumption of natural resources like water, air, etc. has been increasing rapidly due to increase in population as well as change in life style. The short term social issues and long term environmental issues are interrelated. It has been found that the impact of environmental degradation is often most severe on the poor and disadvantaged segments of the society. Use of available resources optimally has therefore emerged as one of the essential prerequisites of sustainable development. This would be possible with introduction of appropriate technology and better management practices. Involvement of the people and the local leadership play a key role in ensuring this. In this context, learning from others' experience and replicating the success stories has special significance not only for expediting the pace of economic development but in making it sustainable as well.

Unemployment and poverty are the major challenges of the developing nations. It is necessary to generate productive employment on a large scale, giving due attention environment and needs of the disadvantaged segments of the society for making the development process socially inclusive. In view of its multifarious advantages, mulberry sericulture is considered as an effective tool for economic empowerment of the disadvantaged segments. Sericulture covers both land based activities relating to plantation as well as industrial activities like reeling, twisting and weaving. Its labour intensive nature provides a solution to the problems of seasonal unemployment and underemployment. Special features of silk fabrics would maintain the demand in national and international markets.

Further as silk fabric is purchased by the rich; it serves as a conduit for transfer of resources from the rich to the poor thereby *ensuring growth with distributive justice*.

As mulberry sericulture requires land and plenty of labour, its expansion depends on the stage of industrialization of any particular area. With industrialization and urbanization, labour becomes costly and land becomes scarce. These restrain expansion and viability of the sericultural operations. It has been noticed that as a result of industrialization and urbanization, mulberry sericulture shifts from the traditional areas to new and nontraditional areas. This happened in several countries across the world. Initially Japan used to be the centre of sericulture activity, which has shifted to China in the last decades. In India, Karnataka and adjoining area of Tamil Nadu used to be the hub of sericulture industry. Following industrialization and urbanization, the centre of silk production is gradually shifting to other areas. However, attempts for introducing sericulture in new areas in India have got only limited success. In this background, exploring the factors responsible for stabilizing sericulture in new and nontraditional areas based on a firsthand experience of initiative taken would be an interesting exercise.

Experiences of Odisha

Experiences of introducing mulberry sericulture on a project mode in Odisha in the Nineties have revealed number of interesting insights. It has confirmed the potential of mulberry sericulture not only in starting in a so called non-traditional area but also with Scheduled Tribe beneficiaries. The practice was taken up under rain-fed condition and up to 3 crops linked up with monsoon could be reared in a year. In spite of specific target and attempts made, it was neither possible to take up plantation under irrigated condition nor to attract progressive farmers for taking up sericulture. As a result, even though the practice was introduced, its sustainability on a long term basis remained limited unlike the traditional states.

So far as return form mulberry sericulture is concerned, mulberry plantation in half a acre of land under rainfed condition along with silkworm rearing generated income varying from Rs 2000 to Rs 3000 per year[2]. While this amount was not adequate enough to enable a family to cross poverty line, it was higher than the income from other land based activity taken up under similar conditions. It appeared to be a good supplementary income and contributed positively in settling down the nomadic families at one place.

As regards employment, about two members of the family got seasonal employment in mulberry plantation and silkworm rearing. This was particularly useful to the women, as they got the work inside the house itself and could do it while attending to other household chores like cooking, washing, cleaning, looking after the children etc. Less strenuous and convenient nature of mulberry sericulture made the practice particularly attractive to the women in rural areas.

On the technical side, high yielding variety of mulberry plant was introduced, which gave better quality and higher quantity of mulberry leaves. However the plantation was raised mainly with mulberry cutting and use of sapling in raising plantation remained limited. Yield of mulberry leaves, quantity of silkworm reared, yield of mulberry cocoon and return form sericulture remained much lower than the target. These were partly due to the limited resource base of the rearers and partly due to other factors such as low soil nutrition, plantation raised under rainfed condition etc. It was interesting that viability of rearing of bivoltine breed of silkworm could be taken up in new areas and the myth "*that it is necessary to initially go for rearing multivoltine breed and then switching over to bivoltine*" was proved to be wrong. In a new area, the limited choice of the rearers and going by the advice of the project authorities was primarily responsible for introducing the bivoltine breed. Unlike the traditional states, there was not much resistance to change from multivoltine to bivoltine. However, the system of chowki rearing could not be established and rearing was done with the silkworm eggs procured from the grainage.

Organization of the rearers into primary mulberry rearers cooperative society at the block level and the primary societies into the apex silk cooperative society at the state level helped in providing the institutional framework for introducing sericulture in general and providing required forward and backward linkages in particular. The primary society took up purchase, collection, stifling, storing of cocoon as well as reeling and twisting operations with help of the apex society, which were of critical importance in dealing with perishable crops and stabilizing mulberry sericulture in a new area. Even though the price for the cocoon paid by the society to the rearers was some what less than the market based price, support price provided to the rearers through the society bridged a vital gap in starting the practice in the area.

On the whole, experience of Odisha indicated that mulberry sericulture could be introduced in new areas on project mode with required linkages. However, there were shortfalls in achieving the desired levels of productivity, financial return and attracting progressive rearers. The striking feature of this initiative related to its social aspect in using mulberry sericulture as a tool for poverty alleviation, particularly in settling down the nomadic tribes at one place.

Experiences of Tripura

The Tripura study was more detailed and recent. Experiences of Tripura are centered on a special Sericulture Project launched in 1995 in selected clusters with exclusively women beneficiaries. The project was supported over the next two decades. Like in case of Odisha, in Tripura the activity was taken up by mobilizing woman rearers majority of belong to Scheduled Tribes, in small land holdings under rainfed condition.

Due to relatively high average rainfall and good soil conditions, it was possible to take up to four silkworm rearings in a year. The practice remained limited restricted to small farmers and there was not much headway in attracting progressive farmers nor providing irrigated land for sericulture. Soil fertility, low organic matter content, lack of irrigation etc. restricted productivity and return from sericulture, at a level lower than that of the traditional area.

Raising mulberry sapling in decentralized nursery and using it for mulberry plantation was a common feature in Tripura. Apart from stabilizing the plantation, it helped in providing supplementary income to the women groups raising the nursery.

Rearing of pure *bivoltine* breed of silkworm by the first generation women belonging to Scheduled Tribe was the most interesting aspect. The farmers were assisted for having separate rearing house and rearing equipment. The yield of cocoon from 100 Dfl and per acre was comparable with that of the traditional states, which was noteworthy.

Dovetailing fund available under employment generation schemes like MGNREGA was another interesting feature in case of Tripura. MGNREGA fund was used for raising and maintaining mulberry plantation and construction of rearing house. Support over a longer term helped in increasing yield of the crop and income of the rearer. In addition to stabilizing sericulture, it led to creation of productive assets out of MGNREGA fund.

Setting up key infrastructure like grainage for production of bivoltine silkworm egg was another significant achievement. The staff and officers of the department were trained and assisted for running the grainage. It was found that they did this work with perfection. Further, the arrangement for procuring, stifling, reeling and twisting was done with in a well coordinated manner. So far as conversion of yarn to fabric is concerned, Tripura's handloom base was limited to narrow width loin looms used by the tribals for weaving cloth for own domestic consumption. In the absence of traditionally strong commercial handloom base (like Odisha), an attempt was made to take up silk weaving with power loom for converting the silk yarn produced into fabrics. The silk fabrics were printed and sold as *saree*, which provided good marketing linkage for the silk yarn. This was yet another innovative and pragmatic step.

Effective use of MGNREGA fund, setting up of required infrastructure for provision of forward and backward linkages were major developments in case of Tripura. However it was not possible to persuade progressive farmers and raise mulberry sericulture under irrigated condition. Thus while mulberry sericulture was introduced and the rearers got return from sericulture, it remained lower than the traditional areas.

Way Forward

As a natural fabric with several unique features, Silk has its own place in the textile industry. Production of manmade fabrics based on new technology has brought several alternatives to silk. However silk has retained its relevance and importance, which would continue in the coming decades. Silk yarn and fabric are produced through natural process involving raising mulberry garden and rearing silkworm. As such, availability of land and cheap labour are essential requirements for sustaining mulberry sericulture. Experience worldwide indicates that as the land become scarce and labour become costlier, sericulture practices shift from one region to another. Initially, Japan was the main producer of mulberry silk. However, following industrial development of Japan, the distinction of being the leading global silk producer shifted to China. In India also mulberry sericulture has been shifting from Karnataka to other states. In due course, it is bound to shift to the relatively less industrialized and urbanized parts of the country in the North East and the Deccan Plateau regions.

In this background, introduction of mulberry sericulture in new area is required to be appreciated. The strength of mulberry sericulture depends upon market potential for silk fabrics. It acts as a medium for transfer of wealth from the rich to the poor, as silk fabrics is purchased by the rich and is produced by the poor, who took up plantation, silk worm rearing, reeling, weaving, etc. The practice is labour intensive in nature and particularly meaningful to the women. Sericulture activities involve cultivation of mulberry, which help in the conservation of soil, water and environment. Thus, mulberry sericulture acts as a tool for solving the problems of economic deprivation along with environmental conservation.

Study of the experience of introducing mulberry sericulture in new and nontraditional areas of Odisha and Tripura has indicated its potentials. Based on the same, it would worthwhile to look into certain specific features, as discussed below, for optimizing use of available resources and sustaining the practice in the new areas.

- The first and foremost thing about introduction of mulberry sericulture in a new area is to appreciate its technical aspects. Undertaking any new practice particularly with the poor has always been a challenge because of their low income and risk taking capacity. It becomes more difficult in case of a perishable commodity. Introduction of sericulture in a new area becomes particularly difficult as unlike milk and vegetables, the produce cannot be consumed locally. Further, all the three components of sericulture namely mulberry leaves, silkworm and silk cocoon, are perishable. Mulberry leaves are prone to cattle grazing, when it is raised in isolated fields. Mulberry silkworm is prone to diseases particularly in hot and humid climatic condition. Mulberry cocoon is also perishable as the moth emerges within 5 to 6 days of spinning unless it is killed by stifling. Introduction of mulberry sericulture in a new area should therefore be taken up only in a coordinated manner with required technical expertise and close supervision. Based on the initial success achieved, at times there are temptations to expand to new area at random. However such tendency is bound to meet with failures and should be avoided at any cost. Mulberry sericulture should be *introduced only in geographically compact clusters.*

- In view of several advantages of mulberry sericulture, there are attempts to use it as a tool for empowering the tribals and often those, who have been nomadic and practicing shifting cultivation (*slash and burn method* of agriculture). Such communities are often the first generation agriculturists and new to settled agriculture. Further the land available with them has low fertility with poor organic content. In such cases, under taking any economic activity is difficult, it is more so in case of mulberry sericulture. It is necessary to provide intensive guidance, adequate support and close monitoring in addition to financial support for making headway with such group of beneficiaries.
- While introducing sericulture in a new area, all families owning land in the identified geographical cluster should be encouraged and assisted for taking up sericulture, instead of limiting it only to rearers from below poverty line (BPL). This would not only facilitate extension and provision of services and support but also mobilize group dynamics and provision of community support like checking cattle grazing.
- While persons belonging to Below Poverty Line (BPL)

may be assisted under various poverty alleviation programmes, the families above poverty line may also be given attention. They may be given required financial support as soft loan and inputs. This would help in induction of entrepreneurial ability, which is essential for success of any new programme.

- Mulberry plantation should be raised only with sapling of high yielding variety, which would give good return and stabilize early.
- Mulberry sapling may be raised through *Decentralized People's Nursery* through self-help groups, which may be further supplemented by private efforts.
- Special attention may be given to improving soil fertility and water retention ability by taking up innovative practices like application of farmyard manure, vermi compost, and lime in acidic soil.
- Appropriate technology like pot irrigation, drip irrigation, mulching for reducing evaporation etc. may be taken up for provision of irrigation facility.
- In case of beneficiary belonging to disadvantaged segments, resource available under wage employment schemes like MGNREGA may be dovetailed for strengthening the mulberry plantation.
- Each rearer should be assisted for having a separate rearing house and required rearing equipment.
- Training and capacity building of the rearers may be taken up intensively and on regular basis.
- Silk worm *Grainage*, with required facility and trained staff, should be set up locally for producing diseases free laying with quality and ensuring supply in time.
- For extension adequate number of staff and officers should be posted and they should be trained at regular intervals.
- Arrangements should be made for purchase, stifling and storing of cocoon as well as reeling and twisting. For the purpose, the rearers may be organized into primary cooperative, which may look after marketing of cocoon. Interested private reelers may also be involved for running the silk reeling and twisting in the private sector.
- Due to decline in number of traditional handloom, there

are often difficulties in converting silk yarn into fabrics. Silk weaving in power-loom should be integrated for overcoming this. In such cases, weaving plain silk fabrics along with printing and processing would fetch a good price and help in making the operation financially viable.

- Future of expansion of sericulture in the country has to be necessarily based on its expansion to new and non-traditional area. The forests land, over which right have been given to the forest dwellers, provide one such scope which needs to be explored. Plantation of tree type mulberry can be taken up in such undulating hilly terrain along with inter cropping. From the study, it appears that mulberry sericulture has potential to provide supplementary income to the poorer section of the society, who has small landholding but surplus labour. Provision of adequate assistance required for taking up sericulture as per technical specification and infrastructure for provision of forward and backward linkage are the basic pre-requisite for taking sericulture to new area. While the infrastructure may be set up on the public sector initially, its management should be transferred to private sector in due course. Besides necessary action has to be taken for improvement of soil fertility, drip irrigation etc. for increasing productivity.

Policy Implications

The present studies provide several policy related implications, some of which may be outlined as follows.

- Mulberry sericulture has the potential of being used as a tool for improving the socio-economic condition of the families living below poverty line in general and the tribals in particular. Study of capital intensive projects with long gestation period like the hydroelectric power plant often cause severe hardship to the locals due to displacement, loss of livelihood, inadequate rehabilitation etc. Local inhabitants feel deprived and marginalized by the outsiders. These get reflected in form of protests, which at times turn violent. Mulberry sericulture is a land based activity similar to the traditional practice and is therefore more acceptable to the tribals. The limitation in financial returns from sericulture is required to be overcome by suitable measures and adoption of appropriate technology. This potential can be exploited for uplifting the poor and the tribals by providing livelihood on a sustainable basis, which would in turn help in achieving the much cherished goal of *growth with distributive justice*.

- Enactment of MGNREGA has been a landmark initiative for empowering the poor and the unskilled/ semiskilled unemployed. While implementation of MGNREGA has been widely appreciated in providing employment and ensuring growth with equality, creation of productive assets under MGNREGA has been a matter of concern. In view of labour intensive nature, good economic return and demand for silk in national and international market, promotion of mulberry sericulture offers an effective means for the empowerment of small and marginal farmers, who have low capital but surplus labour. The practice is also gender sensitive as silkworm rearing does not require hard and strenuous work and is usually done by women indoor along with other household works. It has both agricultural and industrial components, which ensure utilization of surplus labour optimally. Dovetailing of MGNREGA fund in taking up mulberry sericulture in preparation of sapling, plantation of mulberry, construction of rearing house, etc has been found to be useful. Such synergy ensures generation of employment along with creation of productive assets. Pursuing these further by taking up drip irrigation for raising mulberry plantation under irrigated condition and improving soil fertility will lead to multiple cropping and higher return to the farmers.
- Promotion of mulberry sericulture also helps in saving valuable foreign exchange. Indian silk industry has been depending upon China for getting quality silk yarn, which is used in power loom and in handloom as warp. Even though production and quality silk has been improving, dependence on import of Chinese silk still

continues. It has been found that it is possible to take up bivoltine sericulture and produce quality silk in new areas. Pursuing this would help in increasing production of quality silk, which will reduce import of silk yarn from China and save foreign exchange.

- Promotion of mulberry sericulture helps in conservation of soil, water and environment. In that sense, it fulfills the twin objectives of sustainable development viz (i) social relating to transfer of resources from the rich to the poor and (ii) environmental relating to conservation of nature.

Mulberry sericulture has the potential of being an effective medium for development of rural areas in general and poverty alleviation in particular. It has several unique features like being labour intensive, of particular relevance to women, higher return than other land based activities throughout the year and checking urbanization by development of rural areas. Mulberry sericulture ensures transfer of resources from the rich, who consumes silk, to the poor, who produces it. These features make it an ideal tool for development, as compared to other capital intensive activities. Productions of quality silk reduce import, save foreign exchange and make the country self-dependent. With a well-coordinated approach and proper strategy for development, mulberry sericulture can ushers in development in the true sense and establishing a socialistic society and ensuring "*Growth with distributive justice*".

Bibliography

Alak Bhattacharya and Valnamuana Darlong, 2013, Paradigm of Human Development of Tribal Areas in Tripura. Indian Journal of Research. 2 (4), 31-324.

Aruga, H., 1994, *Principles of Sericulture*. Oxford and IBH Pub. Co. Pvt. Ltd., New Delhi.

Basavaraja, H.K.; Nirmal Kumar, S.; Suresh Kumar, N.; Mar Reddy, N.; Giridhar, K.; Ahasn, M.M. and Datta, R.K. 1995, New productive bivoltine hybrids. *Indian Silk*, **34**(2): 5-9.

Baskar H., 2005, Indian Sericulture: Exciting Path Ahead International Sericultural Commission, India.

Brahma, K.C., B.D. Dash, S.C. Nayak, S.K. Parichha and D. Chakravorty, 2011, Socio economic development of tribal's through tasar culture in Odisha.

Dandin, S.B.; Jaiswal, J. and Giridhar, K. 2010. A Hand Book of Sericulture, Central Silk Board.

Data, R. K. 1984. Improvement of silkworm races *Bombyx mori* Sericologia, 34(3): 393-415.

Datta, R.K. 2002. Mulberry cultivation and utilization in India. FAO proceedings of an electronic Conference. Ed. Sanchez, M. 2002.

Datta, R.K. and Nanavaty, M. 2005. Global Silk Industry—A complete Source Book. Universal Publishers, USA. www.universal-publishers.com.

Dewangan S.K., Sahu K.R. and Soni S.K., 2011, International Science Congress Association 371 Vol. 1 (ISC-2011), 371-374 (2012) Res. J. Recent. Sci.

Dewangan, S.K., 2010, Sericulture—A Tool of Eco-System

Checking Through Tribal, Journal of Environmental Research and Development, Vol. 6 No. 1, 31-37.

Directorate of Handloom, Handicrafts and Sericulture 2011. Promotion of mulberry sericulture under MGNREGA experience of Tripura. Published by the Government of Tripura.

Dwarakinath, R. 1994. Guide to Sericulture Extension. Central Silk Board, Bangalore.

Gangopadhyay D. 2009. Sericulture Industry in India—A Review, NISTDS, CSIR, New Delhi.

http://cultivation-information.blogspot.in/2011/11/sericulture-in-tripura-sidelined.html.

http://en.wikipedia.org/wiki/Economy_of_India.

http://en.wikipedia.org/wiki/Economy_of_India#Income_and_consumption.

http://en.wikipedia.org/wiki/Ocupy_Wall_Street.

http://en.wikipedia.org/wiki/Sustainable_development.

http://nrmcindia.co.in/news.php.

http://planningcommission.nic.in/. Tenth Five Year Plan, 2002-2007, Planning Commission, Government of India: 717

http://planningcommission.nic.in/plans/stateplan/sdr_pdf/tripura%20hdr.pdf. Tripura Human Development Report 2007, Government of Tripura, Tripura.

http://rural.tripura.gov.in/. Achievement Report 2007-2012, Rural Development Department, Government of Tripura, Tripura.

http://texmin.nic.in/policy/Fibre_Policy_Final report of Sub-Group on Silk-National Fibre Policy, Ministry of Textiles, Government of India.

http://trak.in/tags/business/2012/03/30/average-per-capita-income-indian-states/

http://www.csb.gov.in/- An official website of Central Silk Board, Ministry of Textile, Government of India.

http://www.destripura.nic.in

Huang Guo Rui. 1998. Silk Reeling Chinese Bok translated and published by Oxford & IBH Publishing company Pvt. Ltd. New Delhi, India.

Katsumata, F. 1975. Silkworm mounting. In: textbook of tropical Sericulture, 3rd Edition, Japan Overseas Co-operation Volunteers, Tokyo Japan.

Khaken Debbarma, 2011, Status of tribal in Tripura State of North-East India-An overview feedback@tiprasa.com).

Kibria, S.A.M.S., 1990, Handbook on pest and disease control of mulberry and silkworm. Published by FAO of The United Nations, Economic and social commission for Asia and the Pacific, Bangkok, Thiland. pp. 3-88.

Mano, Y.; Nirmal Kumar, S.; Basavaraja, H. K.; Mal Reddy, N. and Datta, R. K. (1993) A new method to select promising silkworm breeds/combination. *Indian Silk*, **31**(**10**): 53.

Masanori, S. and Yatro, T. (1972) Hand book of silkworm rearing. Fuji publication Co Ltd. Japan.

MGNREGA Annual Report 2011-12, Gomati District, Government of Tripura, Tripura.

MGNREGA Outlays and outcome-2011-12, Sepahijala District, Government of Tripura, Tripura.

Mishra, S.N. 2004. Indian Economy and Socio-Economic transformatrion, Orissa Economic Association. Deep & Deep Publication Pvt. Std. New Delhi.

Naik, G and Babu, K. R. 1993. Demand and supply prospects for high quality raw silk. Oxford & IBH Publishing company Pvt. Ltd. New Delhi, India.

Nakamura, A. 2000. Oxford & IBH Publishing company Pvt. Ltd. New Delhi, India.

Nanavathy, M. M. 1990, Silk production and processing and marketing, Willey Eastern Limited, New Delhi : 1.

Nanavathy, M. M. 1990, Silk production and processing and marketing, Willey Eastern Limited, New Delhi : 1.

Nina Hyde, 1984, The Queen of Textiles. National Geographic Magazine. Vol. 165, No. 1, January, pp. 2-49.

Rajpurohit, A.R. and Govindraju, K.V. 1980. A study of income and employment in sericulture, Shiny Publication.

Report of the Indian Tariff Board, 1933. Tariff Commission Ministry of Commerce and Industry Government of India, New Delhi, INDIA.

Royle, J.F., 1840, Essay on the productive resources of India, M.H. Allen & Co. London.

Saratchandra B., 2005, Strides in Indian Sericultural Science and Technology International Sericultural Commission, India.

Shahid Yusuf, 2001. Facts of Globalisation-International and local dimensions of development. The World Bank.

Silk Review, 1997. A survey of International Trends in production and trade. ITC, UNCTAD/GATT Publications.

Singh, C.P. 1989. Poverty aleviation programme under plans. Indus Publishing Company, New Delhi

Sinha, S. 1990. Development of Indian Silk- a wealth of oportunity. Oxford & IBH Publishing Company Pvt. Ltd. New Delhi, India.

Sonwalkar, T.N. 1991. Handbook of Silk Technology. Willey Eastern Limited, New Delhi : 1.

The Cambridge Economic History of India, Vol. I & II 1984. Publisher Orient Longman, 1984.

Ullal, S.R. and Narasimhanna, M. N. 1987. Handbook of practical Sericulture. Bangalore, Central Silk Board.

Uttar Pradesh District Gazetteers Varanasi, 1962, Government of Uttar Pradesh.

Index